SpringerBriefs in Energy

More information about this series at http://www.springer.com/series/8903

Weidong He · Weiqiang Lv
James H. Dickerson

Gas Transport in Solid Oxide Fuel Cells

 Springer

Weidong He
Weiqiang Lv
School of Energy Science and Engineering
University of Electronic Science
    and Technology of China
Chengdu
China

James H. Dickerson
Department of Physics
Brown University
Providence, RI
USA

ISSN 2191-5520   ISSN 2191-5539   (electronic)
ISBN 978-3-319-09736-7   ISBN 978-3-319-09737-4   (eBook)
DOI 10.1007/978-3-319-09737-4

Library of Congress Control Number: 2014946743

Springer Cham Heidelberg New York Dordrecht London

Printed on acid-free paper

Springer is part of Springer Science+Business Media (www.springer.com)

Preface

The ultimate goal of this book is to provide an integrated view of the basic theory, materials science, and engineering of gas transport in solid oxide fuel cells (SOFCs). Further, this book will provide an invaluable, contemporary reference for the development of fundamental theory and experiment, advanced experimental measurement techniques, and industrial applications of gas diffusivity in solid oxide fuel cells.

Interest in fuel cell technologies has been motivated by their function: directly converting stored chemical energy into electrical energy without combustion and emission of pollutants, such as nitrogen oxides (N_xO_y). These devices can overcome combustion efficiency limitations since the operation of fuel cells does not necessarily involve the Carnot cycle, thus reducing the emission of pollutants. Compared with other types of fuel cells, solid oxide fuel cells have shown clear advantages over other systems, since hydrogen, hydrocarbons, carbon monoxide, and carbon can be utilized as constituent fuels. The major disadvantage of SOFCs is their high operation temperature, which can reach 1000 °C. At such high temperatures, few materials can function effectively as electrolytes or electrodes. This feature of SOFCs increases their operation and fabrication costs, and hinders their application in rapidly developing areas of application, such as in portable power and automobile power device applications. The impedance of SOFCs, including the activation and concentration polarizations of electrodes and the Ohmic loss of electrolytes, increases sharply with decreasing operating temperatures. To reduce the impedance, fundamental comprehension of the mechanism of gas diffusion through the electrode and that of gas transport between the electrode and the electrolyte is necessary. Mechanisms and mathematical models of gas diffusion are discussed in detail in the first chapter of this book.

Several techniques for directly measuring gaseous diffusivity have been developed in recent years. These techniques allow gas transport coefficients to be accurately evaluated. The results of these measurements help to optimize the configuration of solid oxide fuel cells, including the surface properties of electrodes and the structure of electrodes and electrolytes, as well as the techniques for preparing electrolytes. Recent theoretical and experimental advancements in these measurement techniques are discussed in the middle chapters of this book.

Gas diffusivity of electrodes in solid oxide fuel cells drops rapidly with reducing operation temperatures. This loss of diffusivity cannot be compensated through the optimization of the configuration of the fuel cell. Therefore, the key to lowering the operation temperature of solid oxide fuel cells is the development of high-efficiency electrodes. The role of gas diffusivity measurement techniques in the exploration of novel electrode materials are also explored in the middle chapters of this book. Then, the book focuses on the strategies of realizing advanced solid oxide fuel cells with improved gas transport. This chapter presents an overview of novel porous electrode materials, and the techniques allowing for the rational design of electrode microstructure with highly efficient gas transport parameters, including porosity, tortuosity, etc. Finally, an outlook on research and development of low-temperature solid oxide fuel cells is presented.

Acknowledgments

The authors are grateful to those who provided generous help and encouragement during the writing of the book. We are grateful to and cordially acknowledge these individuals, a limited number of whom are listed as follows:

Prof. Qi Huang, School of Energy Science and Engineering, University of Electronic Science and Technology of China, PR China;
Ms. Jiangwei Li, Department of Literature and Journalism, Sichuan University, PR China;
Ms. Yinghua Niu, School of Chemical Engineering, Harbin Institute of Technology, PR China;
Mr. Chengyun Yang, Heolo Technology Corporation, PR China;
Prof. John B. Goodenough, Texas Material Institute and Materials Science and Engineering Program, University of Texas at Austin, US;
Dr. Kelvin HL Zhang, University of Oxford, UK and Pacific Northwest National Laboratory, US;
Dr. Wayne P. Hess, Pacific Northwest National Laboratory, US;
Dr. Zhenjun Li, Pacific Northwest National Laboratory, US;
Dr. Jungwon Park, UC Berkeley and Harvard University, US;
Dr. Bin Wang, Vanderbilt University and University of Oklahoma, US;
Dr. Junhao Lin, Vanderbilt University and Oak Ridge National Laboratory, US.

Contents

Nomenclature

E	Potential (V)
E^o	Equilibrium redox potential (V)
E_{ocp}	Open circuit potential (V)
T	Absolute temperature (K)
η_{ohm}	Voltage drops due to ohmic polarization losses (V)
η_{act}	Voltage drops due to activation, losses (V)
η_{con}	Voltage drops due to concentration polarization losses (V)
K_n	Knudsen number
d_p	Diameter of the pore (cm)
d_g	Effective diameter of a gas molecule (cm)
λ	Gas mean free path (cm)
k_B	Boltzmann constant (1.3807×10^{-23} J/K)
P	Gas pressure (Pa)
P_t	Total gas pressure (Pa)
D_i	Bulk diffusivity of gas species i (cm^2/s)
D_{ij}	Binary diffusivity of gas i and j (cm^2/s)
D_i^t	Total diffusivity of species i (cm^2/s)
D_{ij}^{eff}	Effective binary diffusivity of gas i and j (cm^2/s)
D_{iK}	Knudsen diffusivity of gas i (cm^2/s)
$D_i^{t,eff}$	Total effective diffusivity (cm^2/s)
ϕ	Porosity
T	Tortuosity
R	Gas constant (8.314 J/(mol K))
c	Total gas molar concentration (mol/L)
c_i	Molar concentration of gas i (mol/L)
μ_i	Chemical potential of species i (J/mol)
μ_0	Standard chemical potential of gas i under 1 atm and 1 mol/L (J/mol)
M_i	Gas molecular weight (g/mol)
Ω	Collision integral
σ_{12}	Collision diameter (angstrom)
k_i	Effective permeability (m^2)
k_0	Absolute permeability (m^2)

μ_g	Gas viscosity (kg/(m·s))
ρ_g	Gas density (kg/m^3)
ω_i	Mass fraction of gas species i
X_i	The mole fraction of gas species i
C	Total concentration (mol/L)
J	Total net gas transport (mol/(m^2 s))
J_i	Molar flux of gas species i (mol/(m^2 s))
J_j^D	Diffusive molar flux of gas species j (mol/(m^2 s))
J_j^T	Total diffusive and advective molar flux (mol/(m^2 s))
$\bar{D}_2$	Simplified diffusivity, (cm^2/s)
$\bar{c}_2$	Simplified molar concentration (mol/L)
$\bar{k}_2$	Simplified permeability (m^2)
F	Faraday constant (96485.3 C/mol)
l_a	Anode thickness (m)
L_c	Cathode thickness (m)
$p_{H_2}^i$	H$_2$ Pressure inside YSZ tube (Pa)
$p_{H_2}^o$	H$_2$ Pressure outside YSZ tube (Pa)
$p_{O_2}^i$	O$_2$ Pressure inside YSZ tube (Pa)
$p_{H_2}^o$	O$_2$ Pressure outside YSZ tube (Pa)
$p_{H_2O}^i$	H$_2$O Pressure inside YSZ tube (Pa)
$p_{H_2O}^o$	H$_2$O Pressure outside YSZ tube (Pa)
Dx	Diffusivities along x electrode direction (cm^2/s)
Dy	Diffusivities along y electrode direction (cm^2/s)
Dz	Diffusivities along z electrode direction (cm^2/s)
D_s	Summed 3D diffusivity (cm^2/s)
R_i	Ohmic resistance (Ω/cm^2)
i_o	Exchange current density (A/m^2)
i_a	Anode limiting current density (A/m^2)
i_c	Cathode current density (A/m^2)
η_a	Anode concentration polarization (V)
η_c	Cathode concentration polarization (V)
τ_a	Anode tortuosity
τ_c	Cathode tortuosity
ϕ_a	Anode porosity
ϕ_c	Cathode porosity
Δi	Current error (A/m^2)
ΔD	Diffusivity error (cm^2/s)
Δi_a	Anode limiting current density error (A/m^2)
ΔT	Temperature error (K)
Δi_c	Cathode current density error (A/m^2)
η_a	Anode concentration polarization error (V)
η_c	Cathode concentration polarization error (V)
D_{A-mix}^{eff}	Effective diffusivity of A in a multicomponet gas mixture (cm^2/s)
A	Electrode area (m^2)
R$_D$	Gas diffusion resistance (Ω/cm^2)

Abbreviations

SOFC	Solid oxide fuel cell
YSZ	Yttria-stabilized zirconia
PEMFC	Proton exchange membrane fuel cell
MCFC	Molten carbonate fuel cell
SHE	Standard hydrogen electrode
AL	Activation loss
OL	Ohmic loss
CP	Concentration polarization
OCP	Open circuit potential
EMF	Electromotive force
FL	Fick's law
ADM	Advective–diffusive model
MSM	Maxwell–Stefan model
DGM	Dusty gas model
BFM	Binary friction model
LCD	Limiting current density
Ni-YSZ	Porous composites of Ni and YSZ
LSM	Sr-doped $LaMnO_3$
GDC	$Ce_{0.9}Gd_{0.1}O_{1.95}$
SDC	$Ce_{0.9}Sm_{0.1}O_{1.95}$
ESB	$Er_{0.4}Bi_{1.6}O_3$
LSGM	$La_{0.9}Sr_{0.1}Ga_{0.8}Mg_{0.2}O_3$
TPB	Triple phase boundary
MIEC	Mixed ionic and electron conductivity
LSM	$La_{0.8}Sr_{0.2}MnO_{3-\delta}$
LSCF	$La_{0.6}Sr_{0.4}Co_{0.2}Fe_{0.8}O_{3-\delta}$
SSC	$Sm_{0.5}Sr_{0.5}CoO_{3-\delta}$
BSCF	BSCF ($Ba_{0.5}Sr_{0.5}Co_{0.8}Fe_{0.2}O_{3-\delta}$)
LT-SOFC	Low temperature Solid oxide fuel cell
TEC	Thermal expansion coefficient
SEM	Scanning electron microscopy
TEM	Transmission electron microscopy

BET	Brunauer–Emmett–Teller method
FIB	Focused ion beam
FIB-SEM	Focused ion beam—scanning electron microscopy
TXM	Transmission X-ray microscopy
RVE	Representative volume element
MPD	Maximum power density
EIS	Electrochemical impedance spectroscopy
STN	Nb-doped $SrTO_3$
ScYSZ	Sc doped YSZ

Chapter 1
Introduction to Gas Transport in Solid Oxide Fuel Cells

Solid oxide fuel cells (SOFCs) produce electricity by oxidizing fuel gases. The biggest characteristic of SOFCs is their high energy conversion efficiency, up to 60–80 % in theory, since the conversion efficiency is not limited by the Carnot cycle due to the absence of combustion in SOFC devices. Other advantages of SOFCs include fuel flexibility, low emission, long-term stability, and relatively low cost. The major challenge associated with SOFCs is their high operating temperatures, typically above 500 °C. For SOFCs to find a large range of applications for electricity generation in the twenty-first century, numerous efforts are needed to lower the operating temperatures and to enhance the practical conversion efficiency at moderate operating temperatures. The key is to improve mass transfer involved in SOFCs. Since SOFCs operate with gaseous fuels and oxidants, gas transport in the porous electrodes largely influences their performance. In this chapter, a brief introduction to SOFCs will be first given. The main issues to be solved, gas transport phenomenon, as well as the scientific problems in this field will then be depicted in later chapters.

1.1 Introduction to SOFCs

1.1.1 Brief History of SOFC Development

Solid oxide fuel cells were originally realized by Nernst for use as a commercial light source, in an effort to replace carbon filament lamps in the 1900s [1]. The device used Nernst mass, which was made of yttria-stabilized zirconia (YSZ), a conductor of oxide ions in the air. This system operated at temperatures from 600 to 1,000 °C. The electrolyte composition, found in Nernst mass, is still the basis of the commonly used electrolyte. In 1937, Baur et al. [2] developed the first solid oxide fuel cells that used materials, such as zirconium, yttrium, cerium, lanthanum, and tungsten; since then, solid O^{2-} conductor-based SOFCs have attracted increasing attention. In the 1940s, a Russian scientist, O.K. Davtyan, added monazite sand to the mixture of sodium carbonate, tungsten trioxide, and soda glass to increase the conductivity and mechanical strength. However, Davtyan's designs could solve

© The Author(s) 2014
W. He et al., *Gas Transport in Solid Oxide Fuel Cells*, SpringerBriefs in Energy,
DOI 10.1007/978-3-319-09737-4_1

the problems including unwanted chemical reactions and short life ratings. In the late 1950s, research on solid oxide technology began to accelerate at the Central Technical Institute in the Hague, Netherlands, Consolidation Coal Company, in Pennsylvania, and General Electric, in Schenectady, New York. A 1959 discussion of fuel cells noted that problems with solid electrolytes included relatively high internal electrical resistance, melting, and short-circuiting due to semiconductivity. Apparently, many researchers began to believe that molten carbonate fuel cells showed more pronounced short-term promise. Nevertheless, the excellent CO tolerance and the long-term stability still draw the attention of researchers whose focus is on the improvement of SOFC performance for space, submarine, and other military applications. More recently, the emerging energy crisis (climbing energy prices, environmental problems, and advances in materials technology) has reinvigorated work on SOFCs. Numerous companies, universities, and research agencies all over the world are now working in this field.

1.1.2 Principles of SOFCs

As the global population and economy continues to expand, much attention is focused on improving the performance of existing energy systems as well as exploring new forms of sustainable energy sources [3–5]. One of the main proposed strategies toward sustainable energy sources is hydrogen-based fuel cells, such as proton exchange membrane fuel cells (PEMFCs), molten carbonate fuel cells (MCFCs), and solid oxide fuel cells (SOFCs). Fuel cells directly convert stored chemical energy of fuels into electrical energy without combustion and, thus, are capable of overcoming the combustion efficiency limitations as imposed by the Carnot cycle. Moreover, fuel cells reduce the emission of pollutants like nitrogen oxides (N_xO_y) and are environmentally friendly. Compared with other types of fuel cells, solid oxide fuel cells exhibit excellent fuel flexibility, since many chemicals, including hydrogen, hydrocarbons, carbon monoxide, and carbon can be utilized as fuels [6–9].

In an SOFC, as shown by the scheme in Fig. 1.1, oxygen molecules diffuse through the cathode and are reduced to oxygen ions at the cathode active layer. These oxygen ions then transport through a solid ion-conductive electrolyte and react with the fuel gas at the anode/electrolyte interface. The driving force of a fuel cell is provided by the anode/cathode reaction as shown in Eq. 1.1 and Eq. 1.2.

$$\text{Anode reaction: } H_2 + O^{2-} - 2e = 2H_2O \tag{1.1}$$

$$\text{Cathode reaction: } \frac{1}{2}O_2 + 2e = O^{2-} \tag{1.2}$$

The overall reaction of an SOFC is shown in Eq. 1.3

$$O_2 + 2H_2 = 2H_2O \ E^0 = 1.229 \text{ V} \text{ versus SHE } (T = 25\,^\circ C) \tag{1.3}$$

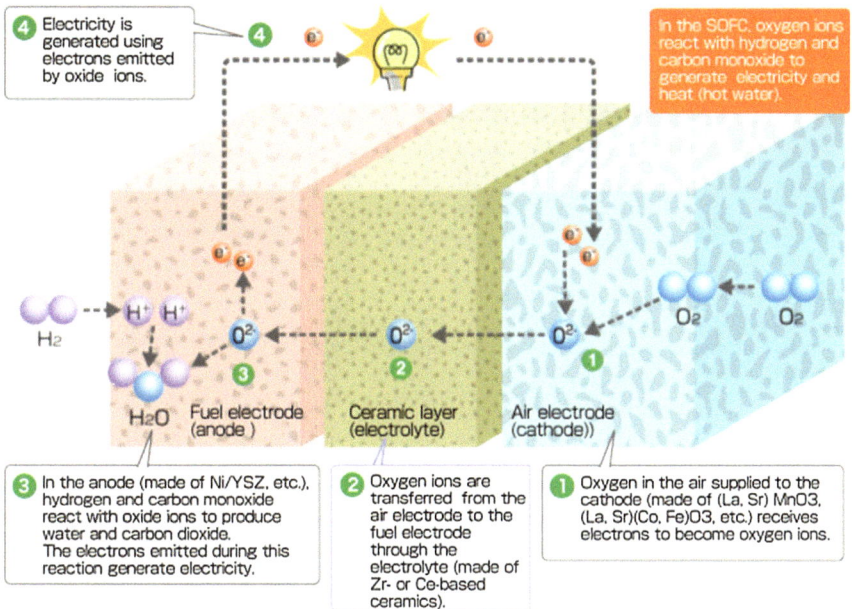

Fig. 1.1 Scheme of the operation principle of an SOFC. SOFCs use a solid oxide electrolyte to conduct negative oxygen ions from the cathode to the anode. The solid oxide electrolyte, usually made of doped ZrO_2 or CeO_2 ceramics, is adequately ironically conductive for O^{2-} only at high temperatures, typically between 500 and 1,000 °C. Adopted from Web site: http://www.osakagas. co.jp/en/rd/fuelcell/sofc/sofc/index.html

The equilibrium potential of the overall reaction E^0 is 1.229 V *versus* SHE at the room temperature (25 °C), which decreases linearly as the temperature increases with a rate of 23 mV per 100 K. For SOFCs operated at 800 °C, the theoretical open potential is ~1.050 V versus SHE [10].

Practical SOFCs are usually fabricated in either planar or tubular structures as shown in Fig. 1.2. Both cell setups exhibit their merits and drawbacks. The advantages of planar cells include construction simplicity, lower fabrication cost, lower ohmic resistance, and higher power density compared with tubular cells. The advantages of tubular cells include no need for high-temperature sealing (the cell can be sealed in the cold ends), and long-term operation stability without obvious decay (the decay rate is only ~0.1 % per 1,000 h). Both designs of SOFCs have been developed extensively by research agencies and industrial communities all over the world.

Solid oxide fuel cells have a large variety of applications ranging from portable power and transport to stationary power supply with outputs from 100 W to 2 MW. Research has led to remarkable success in reducing the energy losses of SOFCs well below 30 %. The major drawback of SOFCs is their high operating temperatures, which are typically above 500 °C. At these temperatures, only a few

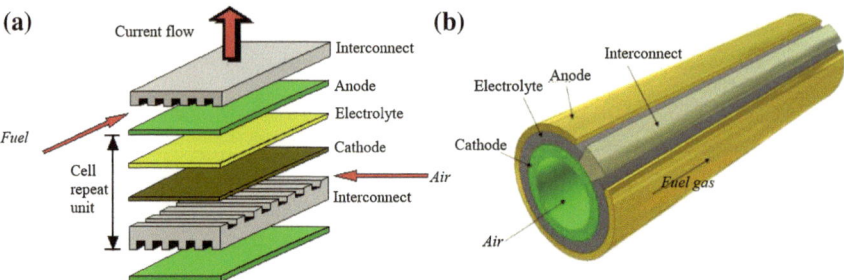

Fig. 1.2 SOFCs with structures of **a** planar cell and **b** tubular cell. Adopted from Web site: http://www.csa.com/discoveryguides/fuecel/overview.php, and http://de.wikipedia.org/wiki/Festoxidbrennstoffzelle

materials can be employed as electrolytes and electrodes. This leads to an increase in cost and hinders potential applications in portable and automobile power sources. Lowering the operation temperatures of SOFCs while maintaining low energy losses has become the key focus in SOFC research [11]. Development of new electrolytes with high ionic conductivities and new electrode materials with high activities at low temperatures (300–600 °C) are paving the way for this direction. In addition to materials development, the rational design of the structure with existing SOFC materials also can greatly decrease various energy polarizations, especially the energy loss caused by gas transport in the operation of SOFCs, and facilitate the operating-temperature-lowering efforts of SOFCs.

1.1.3 Energy Losses in SOFCs

For a gas-based SOFC to operate, the activation energy of the anode/cathode reaction must be overcome [12]. In the process of producing power, the pressure gradient of anode/cathode gas and the concentration gradient of oxygen ions across the anode/cathode and the electrolyte are determined by the transport rates of these gaseous and ionic species. While providing electricity, electrical resistance also is present in all operating components of the fuel cell [13]. These energy losses associated with a fuel cell are divided into three types: activation loss (AL), concentration polarization (CP), and ohmic loss (OL). Figure 1.3 demonstrates the correlation between concentration polarization loss, ohmic loss, and activation loss with different components of a fuel cell. Ohmic loss, which results from electrical resistance, is present across all fuel cell components, whereas activation loss is induced by offsetting energy barrier for catalytic reactions at electrode/electrolyte interfaces. Concentration polarization is induced by the pressure gradient due to limited transport rates of gaseous reactant and/or product species through SOFC electrodes [14].

The cell voltage, E, can be calculated by Eq. 1.4,

$$E = E_{\mathrm{ocp}} - \eta_{\mathrm{ohm}} - \eta_{\mathrm{act}} - \eta_{\mathrm{con}} \qquad (1.4)$$

Fig. 1.3 Energy losses associated with SOFC components. Reprinted from Ref. [15]. Copyright (2013), with permission from Elsevier

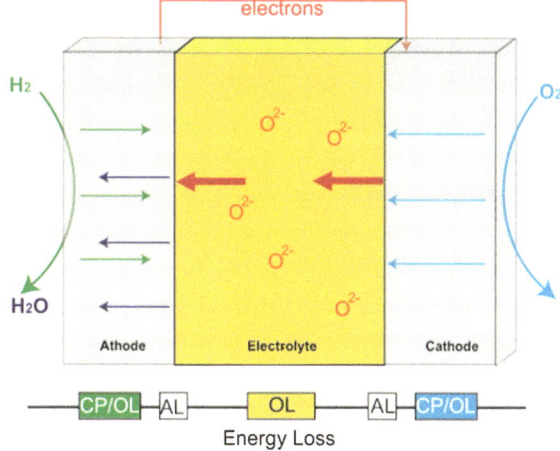

Fig. 1.4 Ideal and actual fuel cell current–voltage characteristics and polarization losses at each region

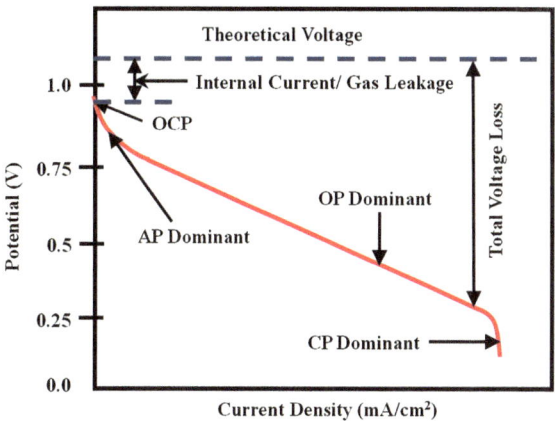

where E_{ocp} is the open circuit potential (OCP) and η_{ohm}, η_{act}, and η_{con} are the voltage drops due to ohmic, activation, and concentration polarization losses, respectively. In principle, E_{ocp} should be the same as the reversible Nernst potential of the reactants, but in many practical cases, the measured OCP is less than the ideal voltage due to leaky gas seals and open pores/micro-cracks in the electrolyte. As shown in Fig. 1.4, for SOFCs working at low output current densities, the activation loss dominates the overall potential drop. The ohmic energy loss increases nearly linearly as the current density increases. At a relatively large output current density, the potential of the cell decreases more rapidly, which indicates the dominance of the concentration polarization. A large output of the current density requires fast consumption of reactant gases and exhaustance of products, exceeding the gas transport rate in the electrodes. Therefore, concentration energy loss should be minimized to increase the energy conversion efficiency and to improve the performance of SOFCs under operation modes of large output current densities.

Porous electrodes are usually employed in SOFCs to improve the gas transport. Factors such as the porosity, pore size, and tortuosity of the electrodes should be optimized to improve the performance of gas transport.

1.2 Gas Transport in SOFCs

1.2.1 General Consideration

In an SOFC system, to produce electricity, continuous supply of fuels and oxidants is required. In the meanwhile, the reaction product must be exhausted outward. The motion of reactant and product species is called mass transport in SOFCs. Although the transport of the oxygen ions in the electrolyte is also a major portion of mass transport, we focus primarily on the mass transport of the uncharged species. The uncharged species are typically gaseous fuels and oxidants, and their transport is gas transport. The motion of the uncharged gases is not affected by the potential gradient and must rely on the diffusion and convection. The gas transport occurs in two areas of SOFC systems—the flow field and the electrode as shown in Fig. 1.5a. The flow field is channels of the size on millimeter-to-centimeter scale, and the gas transport in this area is controlled by convection as driven by the external pressure of the gas sources. On the contrary, the electrodes of SOFCs are porous structures with the pore size on nanometer-to-micrometer scale, and

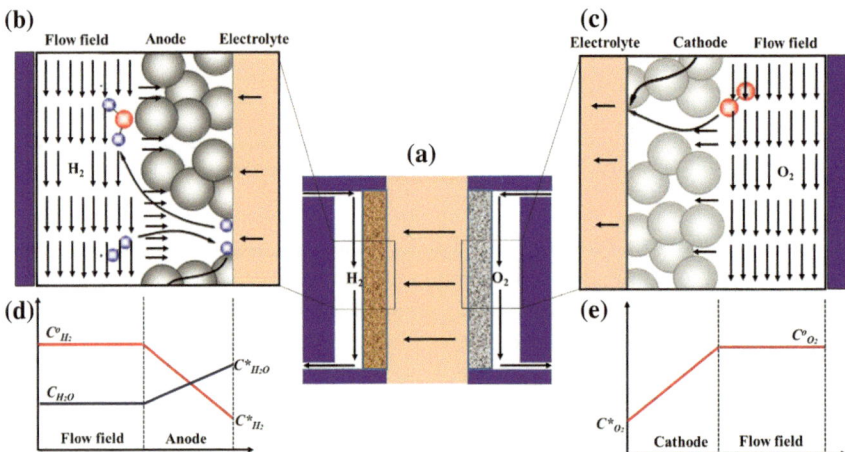

Fig. 1.5 The scheme of gas transport in **a** the H_2–O_2 SOFC system and the diffusion layer, **b** on the anode surface, and **c** on the cathode surface. The consumption of $H_2(O_2)$ at the anode (cathode)–electrolyte interfaces results in the depletion of $H_2(O_2)$ in the anode (cathode). The concentration of $H_2(O_2)$ decreases from the bulk concentration to a lower concentration, as shown in **d–e**. The exhaust of the product of H_2O at the anode is also shown in **d**. At the flow field/anode interface (cathode), the flow rate of $H_2(O_2)$ decreases to zero, which indicates the start point of the diffusion layer

the gas transport in the electrode is mainly controlled by diffusion, as shown in Fig. 1.5b, c. The concentration of $H_2(O_2)$ decreases from the bulk concentration to a lower concentration as shown in Fig. 1.5d, e. The exhaust of the product of H_2O at the anode is also shown in Fig. 1.5d. At the flow field/anode interface (cathode), the flow rate of $H_2(O_2)$ decreases to zero, which indicates the start point of the diffusion layer. The gas transport in the flow field has been discussed elsewhere [16] and is not a concern in this book. The gas diffusion in the electrode will be illustrated thoroughly in the following chapters.

As discussed in Sect. 1.1.3, the reason why we mainly focus on the gas transport in SOFCs exists because the dynamic depletion of the reactants and the accumulation of the product are harmful to the electrode reactions at the active sites. Thus, poor gas transport causes serious concentration energy losses in the cells.

1.2.2 The Driving Force of Gas Diffusion in Electrodes—Concentration Gradient

Gas convection in the flow field results in an adequate mixing of the gas species, and no concentration gradient forms in the flow field. The gas flow rate slows down to zero at the flow field/electrode interfaces and in the electrodes, but the consumption of the fuel or oxidant gases at the electrode/electrolyte interfaces leads to a low concentration of the reactant gas species. These two synergetic effects result in the formation of the concentration gradient of the reactant gases. On the other side, the product of the reactant is continuously generated at the electrode/electrolyte interfaces, and an opposite concentration gradient across the electrodes is formed. The concentration gradients are the driving forces for the diffusion of both reactant and product gaseous species.

1.2.3 Gas Transport in the Porous Electrodes

Porous electrodes are used in fuel cells to maximize the interfacial area of the catalyst per unit geometric area. In SOFCs, an electrolytic species and a dissolved gas react on a supported catalyst. Thus, the electrode must be designed to maximize the available catalytic area while minimizing the resistances for efficient mass transport of gaseous reactants and products. The behavior of porous electrodes is inherently more complicated than that of planar electrodes due to the intimate contact between the solid and gas phases [17]. The gas diffusion in a porous media is significantly different from that in a free space. In macro-channels, as shown in Fig. 1.6a, the gas molecules diffuse freely in the inner spaces since only a small portion of molecules near the walls are blocked due to their collisions with the walls. The gas diffusion in macro-channels follows Fick's law (FL). As gas species diffuse in a porous media with channels or pores on nano-to-micro-scales,

(a) 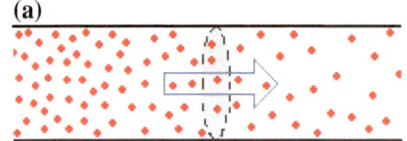   **(b)**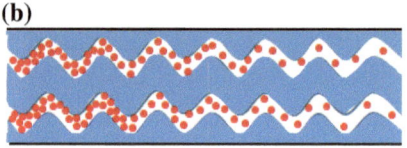

Fig. 1.6 Gas diffusion **a** in free spaces and **b** in the porous media. Figure **a** is adapted from Web site: http://www.chem.uoa.gr/applets/AppletDiffus/Appl_Diffus2.html

as shown in Fig. 1.6b, the portion of molecules near the pore walls increases dramatically. The interaction of the gas molecules with the walls can no longer be ignored. The tortuous path of these pores or channels also increases notably the diffusion length of the gas transport. The gas transport mechanism in the porous electrodes, the measurement techniques, and the correlations associated with the electrode microstructures, the gas diffusion, and the SOFC performance will be the key subjects of this book.

References

1. W. Nernst, W. Wild. Zeitshrift fur Elektrochemie, 1900-12-20
2. Baur, Emil, H. Preis, Zeitshrift fur Elektrochemie (1937)
3. J. Santamaria, J. Garcia-Barriocanal, A. Rivera-Calzada, M. Varela, Z. Sefrioui, E. Iborra, C. Leon, S.J. Pennycook, Science **321**, 676–680 (2008)
4. B.Z. Tian, X.L. Zheng, T.J. Kempa, Y. Fang, N.F. Yu, G.H. Yu et al., Nature **449**, 885–888 (2007)
5. I. Gur, N.A. Fromer, M.L. Geier, A.P. Alivisatos, Science **310**, 462–465 (2005)
6. L. Yang, S. Wang, K. Blinn, M. Liu, Z. Liu, Z. Cheng, M. Liu, Science **326**, 126 (2009)
7. Y.B. Kim, T.P. Holme, T.M. Gür, F.B. Prinz, Adv. Funct. Mater. **21**, 4684–4690 (2011)
8. L.L. Zhang, T.M. He, J. Power Sources **196**, 8352–8359 (2011)
9. T. Suzuki, H. Zahir, F. Yoshihiro et al., Science **325**, 852–855 (2009)
10. R. O'Hayre, A.W. Cha, W. Colella, F.B. Prinz, *Fuel Cell Fundamentals* (Wiley, New York, 2006)
11. E.D. Wachsman, K.T. Lee, Lowering the temperature of solid oxide fuel cells. Science **334**, 935–939 (2011)
12. R.M. Ormerod, Chem. Soc. Rev. **32**, 17–28 (2003)
13. S. Tao, J.T.S. Irvine, Nat. Mater. **2**, 320–323 (2003)
14. F. Zhao, A.V. Virkar, J. Power Sources **141**, 79–95 (2005)
15. W. He, J. Zou, B. Wang, S. Vilayurganapathy, M. Zhou, X. Lin, K.H.L. Zhang, J. Lin, P. Xu, J.H. Dickerson, J. Power Sources **237**, 64–73 (2013)
16. A. Weber, R. Darling, J. Meyers, J. Newman, *Mass transfer at two-phase and three-phase Interfaces*. Handbook of Fuel Cells—Fundamentals, Technology and Applications (Wiley, New York, 2010)
17. C.K. Ho, S.W. Webb, *Gas transport in porous media* (Springer, Dordrecht, 2006)

Chapter 2
Gas Diffusion Mechanisms and Models

2.1 Gas Diffusion in Porous Media

2.1.1 General Consideration

The one-dimensional diffusion of gas molecules in porous media involves molecular interactions between gas molecules as well as collisions between gas molecules and the porous media [1–3]. As gas fuel molecules travel through the porous media, one of three mechanisms can occur, depending on the characteristic of the diffusing gas species and the intrinsic microstructure of the porous media.[3] The three mechanisms are molecular diffusion, viscous diffusion, and Knudsen diffusion. To distinguish among the three mechanisms, the Knudsen number (K_n), which is the ratio of the gas mean free path to the pore size of the electrode, is typically used, as shown in Eq. 2.1 [4]

$$K_n = \frac{\lambda}{d_p} \tag{2.1}$$

where d_p is the diameter of the pores, and λ is the gas mean free path, which can be calculated by Eq. 2.2,

$$\lambda = \frac{k_B T}{\sqrt{2} p \pi d_g^2} \tag{2.2}$$

where P is the gas pressure, d_g is the effective diameter of a gas molecule, k_B is the Boltzmann constant (1.3807×10^{-23} J/K), and T is the temperature of the gas (K). The effective molecular diameters can be estimated using the appropriate covalent and van der Waals radii, while the characteristic or equivalent pore diameter d_p should be evaluated based on the average pore size or chord length distribution [5–8].

   If K_n is much greater than 10, collisions between gas molecules and the porous electrode are more dominant than the collisions between gas molecules, resulting in negligible molecular diffusion and viscous diffusion. If K_n is much smaller than 0.1, collisions and interactions between gas molecules become dominant, and

© The Author(s) 2014   9
W. He et al., *Gas Transport in Solid Oxide Fuel Cells*, SpringerBriefs in Energy,
DOI 10.1007/978-3-319-09737-4_2

Knudsen diffusion becomes negligible compared with molecular diffusion and viscous diffusion. As K_n of a system ranges between 0.1 and 10, all three mechanisms govern gas transport. Different mathematical models have been developed to study the correlation among the parameters associated with the three different mechanisms, such as diffusion coefficient, gas flux, and gas concentration, among others [9, 10]. In this book, diffusion coefficient models are mainly employed.

2.1.2 Molecular Diffusion

Molecular diffusion or continuum diffusion refers to the relative motion of different gas species; this mechanism governs the total diffusion process, as the mean free path of gas molecules is at least one order larger than the pore diameter of the porous media. Fick's law (FL) is the most popular approach to evaluate gas diffusion in clear fluids and gases (non-porous media) due to its simplicity. FL actually has two forms. Fick's first law describes the correlation between the diffusive flux of a gas component and the concentration gradient under steady-state conditions. Fick's second law relates the unsteady diffusive flux to concentration gradient. Fick's first law is depicted in Eq. 2.3 [11]

$$J_i = \frac{-D_i\, c_i \partial(\mu_i)}{RT}\frac{}{\partial x} \quad (i = 1, 2, \ldots, n) \tag{2.3}$$

where J_i is the flux of gas species i, D_i is the bulk diffusivity, R is gas constant, T is temperature, x is one-dimensional diffusion path, c_i is molar fraction of gas species i, and μ_i is the chemical potential of species i at a given state [12]. μ_i is the function of the concentration/density of the mass species and can be expressed by Eq. 2.4,

$$\mu_i = \mu_0 + RT \ln c_i \tag{2.4}$$

where μ_0 is the standard chemical potential of gas i under 1 atm and 1 mol/L.

Fick's second law of diffusion for clear fluids predicts the effects of the concentration change with time on diffusion mechanism, as given by Eq. 2.5.

$$\frac{\partial c_i}{\partial t} = D_i \frac{\partial^2 c_i}{\partial x^2} \tag{2.5}$$

Since we mainly focus on the steady state of SOFCs under continuous operation, Fick's second law will not be discussed in the following section.

The above forms of FL are appropriate for clear fluids or gases. For application in porous media, Fick's first law is often modified by the introduction of a porous media factors, as shown in Eq. 2.6,

$$D_{ij}^{\text{eff}} = \frac{\phi}{\tau} D_{ij} \tag{2.6}$$

where D_{ij} is the binary diffusivity of gas species 1 and 2, D_{ij}^{eff} is the effective binary diffusivity of gas species 1 and 2, and ϕ and τ are the porosity and

Table 2.1 Lennard-Jones potential parameters found from viscosities for the common gas species of fuel cells

Substances		σ (A)	ε/k_B (K)
H_2	Hydrogen	2.827	59.7
O_2	Oxygen	3.467	106.7
N_2	Nitrogen	3.789	71.4
H_2O	Water	2.641	809.1
CH_4	Methane	3.758	148.6
Air	Air	3.711	78.6
CH_3OH	Methanol	3.626	481.8
C_2H_5OH	Ethanol	4.530	362.6

Source Data from [15]

tortuosity of the porous media, respectively. The binary diffusivity D_{ij} can be estimated from the Chapman–Enskog theory, as shown in Eq. 2.7 [13]

$$D_{ij} = \frac{0.00186T^{\frac{3}{2}}}{p\sigma_{ij}^2\Omega}\left(\frac{1}{M_i}+\frac{1}{M_j}\right)^{\frac{1}{2}} \tag{2.7}$$

where D_{ij} is the diffusion coefficient measured in cm²/s, T is the absolute temperature in Kelvin, p is the pressure in atmospheres, and M_i are the molecular weights. The quantities σ_{ij} and Ω are molecular property characteristics of the detailed theory [14]. The collision diameter σ_{ij} given in angstroms is the arithmetic average of the two species as shown in Eq. 2.8.

$$\sigma_{ij} = \frac{1}{2}(\sigma_{i+}\sigma_j) \tag{2.8}$$

The dimensionless quantity Ω is more complex, but typically on the order of 1. Its detailed calculation depends on the integration of the interaction between the two species. This interaction is most frequently described by the Lennard-Jones potential (in 12–6 form). The resulting integral varies with the temperature and the interaction energy. This energy ε_{ij} is the geometric average of contributions from the two species given in Eq. 2.9.

$$\varepsilon_{ij} = \sqrt{\varepsilon_i\varepsilon_j} \tag{2.9}$$

Values of the ε/k_B are given in Table 2.1. Once ε_{ij} as a function of k_BT/ε_{ij} is known, Ω can be calculated using the values in Table 2.2. The calculation of the diffusivities now becomes straightforward as σ_i and ε_i are known.

2.1.3 Knudsen Diffusion

As discussed earlier in this book, when the mean free path of gas molecules is on the same order as the tube dimensions, free-molecule diffusion (i.e. Knudsen diffusion) becomes important. Due to the influence of walls, Knudsen diffusion

Table 2.2 The collision integral Ω with respect to $k_B T / \varepsilon_{ij}$

$k_B T / \varepsilon_{ij}$	Ω	$k_B T / \varepsilon_{ij}$	Ω	$k_B T / \varepsilon_{ij}$	Ω
0.30	2.662	1.65	1.153	4.0	0.8836
0.40	2.318	1.75	1.128	4.2	0.8740
0.50	2.066	1.85	1.105	4.4	0.8652
0.60	1.877	1.95	1.084	4.6	0.8568
0.70	1.729	2.1	1.057	4.8	0.8492
0.80	1.612	2.3	1.026	5.0	0.8422
0.90	1.517	2.5	0.9996	7	0.7896
1.00	1.439	2.7	0.9770	9	0.7556
1.10	1.375	2.9	0.9576	20	0.6640
1.30	1.273	3.3	0.9256	60	0.5596
1.50	1.198	3.7	0.8998	100	0.5130
1.60	1.167	3.9	0.8888	300	0.4360

Source Data from [15]

includes the effect of the porous medium. The molecular flux of gas i due to Knudsen diffusion is given by the general diffusion equation (Eq. 2.10), [16]

$$J_{iK} = -D_{iK} \frac{\partial c_i}{\partial x} \tag{2.10}$$

where D_{iK} is the Knudsen diffusivity. The Knudsen diffusivity of gas species i can be estimated in Eq. 2.11, [17]

$$D_{iK} = \frac{d_p}{3} \sqrt{\frac{8RT}{\pi M_i}} \tag{2.11}$$

where M_i represents the molecular weights of gas species i, and d_p is the mean pore size of the porous media. D_{iK} can be simplified further as Eq. 2.12.

$$D_{iK} = 4850 \, d_p \sqrt{\frac{T}{M_i}} \tag{2.12}$$

In Eq. 2.12, d_p has the unit of cm, M_i has the unit of g/mol, and temperature T has the unit of K.

We can compare D_{iK} with the binary gas phase diffusivity, D_{ij}. First, D_{iK} is not a function of absolute pressure p or of the presence of species B in the binary gas mixture. Second, the temperature dependence for the Knudsen diffusivity is $D_{iK} \alpha T^{1/2}$ versus $D_{ij} \alpha T^{3/2}$ for the binary gas phase diffusivity.

Generally, the Knudsen process is significant only at low pressures and small pore diameters. However, instances exist where both Knudsen diffusion and molecular diffusion (D_{ij}) are important. If we consider that Knudsen diffusion and molecular diffusion compete with one another by a "resistances in series" approach, then the total diffusivity of species i in a binary mixture of i and j, D_i^t, is determined by Eq. 2.13. [18]

$$\frac{1}{D_i^t} \cong \frac{1}{D_{ij}} + \frac{1}{D_{iK}} \tag{2.13}$$

The above relationships for the effective diffusion coefficient are based on diffusion within straight and cylindrical pores aligned in a parallel array. However, in most porous materials, pores of various diameters are twisted and interconnected with one another, and the path for diffusion of the gas molecules within the pores is "tortuous." For these materials, if an average pore diameter is assumed, reasonable approximation for the effective diffusion coefficient in random pores is given by Eq. 2.14.

$$D_i^{t,\text{eff}} = \frac{\phi}{\tau} D_i^t \tag{2.14}$$

The four possible types of pore diffusion are illustrated in Fig. 2.1, with each featured with their respective diffusivity correlation. The first three, pure molecular diffusion, pure Knudsen diffusion, and Knudsen and molecular combined diffusion, are based on diffusion within straight and cylindrical pores that are aligned in parallel array. The fourth involves diffusion via "tortuous paths" that exist within the compacted solid.

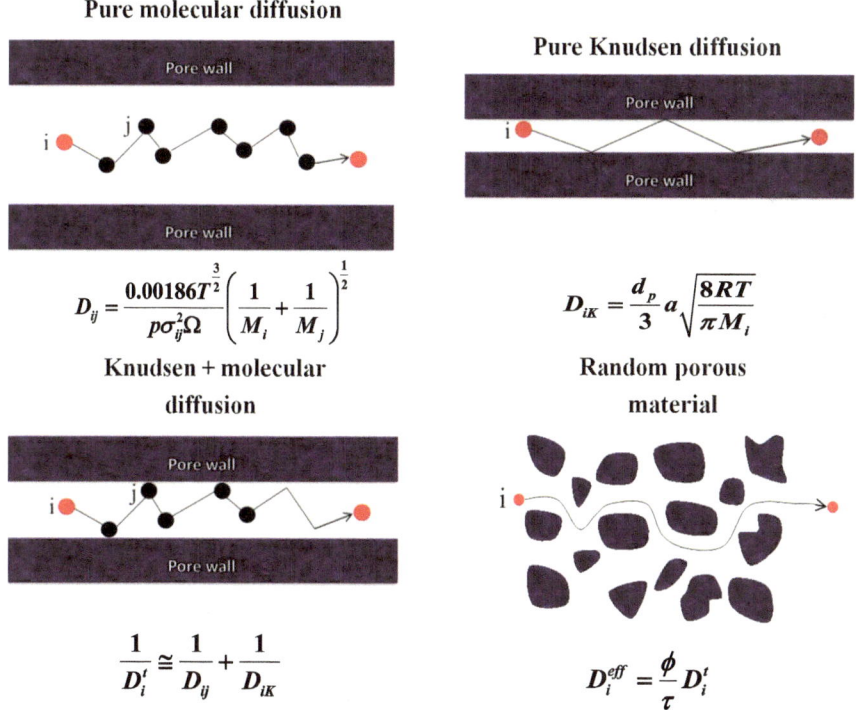

Fig. 2.1   Types of porous diffusion. *Shaded* areas represent non-porous solids. Adopted from Ref. [19]

2.2 Gas Diffusion in Porous Electrodes of Solid Oxide Fuel Cells

2.2.1 Advective–Diffusive Model

In SOFC systems, the electrode pores vary from a few nanometers to several micrometers in size, while the mean free path of fuel gas molecules is a few hundred nanometers. Therefore, K_n is in the range of 0.1–5, and all the three gas transport mechanisms must be considered. Further, the different models must be comprehended to understand the gas transport of a typical SOFC system. Different theoretical models have been proposed to describe the gas transport of SOFC systems. Owing to its inherent simplicity, FL has been most widely employed diffusion-based model. FL considers only molecular diffusion and assumes that the gas flux is proportional to gas pressure gradient. This drawback is addressed in the extended FL model, which combines molecular diffusion modeled by FL and viscous diffusion modeled by Darcy's law. The extended FL, also known as an advective–diffusive model (ADM), is expressed in Eq. 2.15,

$$J_i = -\frac{k_i}{\mu_g}\rho_g\omega_i\nabla p_t - D_{ij}\rho_g\nabla\omega_i \qquad (2.15)$$

where k_i is effective permeability, μ_g is gas viscosity, ρ_g is gas density, D_{ij} is binary gas diffusivity, ω_i is mass fraction of gas species i, and P_t is total gas pressure [20].

2.2.2 Maxwell–Stefan Model

Both FL and ADM only take into account unidirectional interactions in simple dilute binary gas systems; neither of them is valid for ternary or concentrated gas systems where molecular interactions cannot be neglected. To model such complicated systems, the Maxwell–Stefan model (MSM) was developed, which is shown in Eqs. 2.16–2.17,

$$\frac{dX_i}{dz} = \sum_{j=1}^{n}\frac{X_iJ_j - X_jJ_i}{cD_{ij}}, \quad i = 1, 2, \ldots, n \qquad (2.16)$$

$$\frac{dX_i}{dz} = \sum_{j=1}^{n}\frac{X_iJ_j - X_jJ_i}{cD_{ij}^{eff}}, \quad i = 1, 2, \ldots, n \qquad (2.17)$$

where X_i is the mole fraction of gas species i, c is total concentration, D_{ij} is Maxwell–Stefan diffusivity, J_i is molar flux of gas species i, and D_{ij}^{eff} is effective gas diffusivity [21]. MSM can model the gas transport in uniform media or in electrostatic fields.

2.2.3 Dusty Gas Model

MSM fails to model gas systems where gas species frequently collide with the porous media. To model such gas systems accurately, a dusty gas model (DGM) was developed, which is shown in Eqs. 2.18–2.19,

$$\sum_{j=1, j \neq i}^{n} \frac{X_i J_j^D - X_j J_i^D}{D_{ij}} - \frac{J_i^D}{D_{iK}} = \frac{p_t \nabla X_i}{RT} + \frac{X_i \nabla p_t}{RT} \tag{2.18}$$

$$\sum_{j=1, j \neq i}^{n} \frac{X_i J_j^T - X_j J_i^T}{D_{ij}} - \frac{J_i^T}{D_{iK}} = \frac{p_t \nabla X_i}{RT} + \left(1 + \frac{k_0 p_t}{D_{iK} \mu_g}\right) \frac{X_i \nabla p_t}{RT} \tag{2.19}$$

where X_i is gas mole fraction, J_j^D is diffusive molar flux of gas species j, J_i^T is total diffusive and advective molar flux, P_t is total gas pressure, R is gas constant, T is temperature, D_{ij} is binary diffusivity in free space, D_{iK} is Knudsen diffusivity of gas species I, and k_0 is the gas permeability [22]. The DGM considers all possible interactions and collisions and exhibits high accuracy while modeling gas fuel transport through porous fuel cell electrodes. However, DGM has not been as widely used as FL due to the complexity in modeling multicomponent gas fuel systems. The accuracy of DGM coupled with the simplicity of FL yielded the recently developed DGM-FL, which is a more practical gas diffusion model. The model is expressed in Eq. 2.20,

$$J_l = J_l^{\text{diffusion}} + J_l^{\text{convection}} = -\bar{D}_2 \nabla \bar{c}_2 - \bar{c}_2 \frac{\bar{k}_2}{\mu} \nabla p_t \tag{2.20}$$

where $\bar{D}_2$, $\bar{c}_2$, and $\bar{k}_2$ are simplified diffusivity, molar concentration, and permeability of gas species, μ is chemical potential, and P_t is total gas pressure [23]. DGM-FL was obtained by deriving DGM in FL form. This approach results in a simple, efficient, and reliable model of a typical gas transport system. To account for both Knudsen and viscous effects simultaneously, a binary friction model (BFM) was derived, as shown in Eq. 2.21,

$$\frac{dp}{dx} = -RTN(D_{iK}^{\text{eff}} + \frac{k_0 p}{\mu_g})^{-1} \tag{2.21}$$

where P is gas pressure, N is net gas transport, D_{iK}^{eff} is Knudsen diffusivity, k_0 is the gas permeability, and μ_g is the viscosity of the gas species [24, 25]. Compared to Darcy's law that only takes viscous effects into account, BFM shows improved accuracy as it is employed to model gas transport phenomena in fuel cells [26].

2.2.4 Effective Gas Diffusion Model

In the discussed models, the term "effective gas diffusivity" was mentioned. We need to look into multicomponent diffusion to comprehend the term in more details. Effective binary gas diffusivity was developed to simplify computations on the diffusion parameters of a multicomponent gas system since the diffusion fluxes in a fuel cell system are often required at various locations and time points [27–29]. In the effective binary diffusion, a multicomponent gas system is approximated as a binary mixture of gas species i and a composite gas species corresponding to all the other gas species in the system. By introducing effective binary gas diffusivity, any multicomponent gas system can be conveniently treated as a binary gas system. As molecular diffusion, viscous diffusion, and Knudsen diffusion are all considered, the flux of an isothermal binary gas system, such as O_2/N_2 and H_2/H_2O, can be modeled by Eqs. 2.22–2.23 [18]

$$J_1 = -D_1^{t,\text{eff}} \nabla c_1 + X_1 \delta_1 J - X_1 \gamma_1 \left(\frac{c k_0}{\mu} \right) \nabla p_t \tag{2.22}$$

$$J_2 = -D_2^{t,\text{eff}} \nabla c_2 + X_2 \delta_2 J - X_2 \gamma_2 \left(\frac{c k_0}{\mu} \right) \nabla p_t \tag{2.23}$$

where

$$\delta_1 = 1 - \gamma_1 = \frac{D_{1K}^{\text{eff}}}{D_{1K}^{\text{eff}} + D_{12}^{\text{eff}}} \quad \text{and} \quad \delta_2 = 1 - \gamma_1 = \frac{D_{2K}^{\text{eff}}}{D_{2K}^{\text{eff}} + D_{12}^{\text{eff}}} \tag{2.24}$$

$$\frac{1}{D_1^{t,\text{eff}}} = \frac{1}{D_{1K}^{\text{eff}}} + \frac{1}{D_{12}^{\text{eff}}} \quad \text{and} \quad \frac{1}{D_2^{t,\text{eff}}} = \frac{1}{D_{2K}^{\text{eff}}} + \frac{1}{D_{12}^{\text{eff}}} \tag{2.25}$$

In Eqs. 2.24–2.25, J_1 and J_2 are the fluxes of gas species 1 and 2, J is the total flux, c_1 and c_2 are the concentrations of gas species 1 and 2, c is the total gas concentration, $X1$ and X_2 are the molar fractions of gas species 1 and 2, μ is viscosity, k_0 is the gas permeability, P_t is total pressure, D_{iK}^{eff} and D_{2K}^{eff} are the effective Knudsen diffusivities of gas species 1 and 2, and D_{12}^{eff} is the effective binary diffusivity.

References

1. A.S. Joshi, A.A. Peracchio, K.N. Grew, W.K.S. Chiu, J. Phys. D Appl. Phys. **40**, 7593–7600 (2007)
2. M. Cannarozzo, A.D. Borghi, P. Costamagna, J. Appl. Electrochem. **38**, 1011–1018 (2008)
3. J.W. Veldsink, G.F. Versteeg, W.P.M.V. Swaaij, R.M.J.V. Damme, Chem. Eng. J. Biochem. Eng. J. **57**, 115–125 (1995)
4. B. Kenney, M. Valdmanis, C. Baker, J.G. Pharoah, K. Karan, J. Power Sources **189**, 1051–1059 (2009)

5. Jinliang Yuan, Bengt Sundén, Int. J. Heat Mass Transf. **69**, 358–374 (2014)
6. K. Jiao, X. Li, Water transport in polymer electrolyte membrane fuel cells. Prog. Energy Combust. **37**, 221–291 (2011)
7. S. Haussener, P. Coray, W. Lipin´ ski, A. Steinfeld, Tomography-based heat and mass transfer characterization of reticulate porous ceramics for high temperature processing. ASME J. Heat Transfer **132**(2), 023305 (2010)
8. A. Berson, H.W. Choi, J.G. Pharoah, Determination of the effective gas diffusivity of a porous composite medium from the three-dimensional reconstruction of its microstructure. Phys. Rev. E **83**, 026310 (2011)
9. R. Suwanwarangkul, E. Croiset, M.W. Fowler, P.L. Douglas, E. Entchev, M.A. Douglas, J. Power Sources **122**, 9–18 (2003)
10. S.W. Webb, K. Pruess, Transp. Porous Media **51**, 327–341 (2003)
11. A. Fick, Ann. Physik. **94**, 59 (1855)
12. R. Jackson, *Transport in Porous Catalysts* (Elsevier, Amsterdam, 1977)
13. J.O. Hirschfelder, R.B. Bird, E.L. Spotz, Chem. Rev. **44**, 205 (1949)
14. E.L. Cussler, *Diffusion: mass Transfer in Fluid Systems* (Cambridge University Press, Cambridge, 1995)
15. J.O. Hirschfelder, C.F. Curtiss, R.B. Bird, *Molecular Theory of Gases and Liquids* (Wiley, New York, 1954)
16. E.A. Mason, A.P. Malinauskas, *Gas Transport in Porous Media.* (Elsevier, Amsterdam, 1983)
17. R.E. Cunningham, R.J.J. Williams, *Diffusion in Gases and Porous Media* (Plenum press, New York, 1980)
18. F. Zhao, T.J. Armstrong, A.V. Virkar, Measurement of O_2–N_2 effective diffusivity in porous media at high temperatures using an electrochemical cell. J. Electrochem. Soc. **150**, A249–A256 (2003)
19. J. Welty, C.E. Wicks, G.L. Rorrer, R.E. Wilson, Fundamentals of Momentum, Heat and Mass Transfer. Wiley-VCH (2007)
20. C.L. Tsai, V.H. Schmidt, J. Power Sources **196**, 692–699 (2011)
21. W. Kong, H. Zhu, Z. Fei, Z. Lin, J. Power Sources **206**, 171–178 (2012)
22. W. Kast, C.R. Hohenthanner, Int. J. Heat Mass Transf. **43**(5), 807–823 (2000)
23. P. Kerkhof, Chem. Eng. J. Biochem. Eng. J. **64**(3), 319–343 (1996)
24. L.M. Pant, S.K. Mitra, M. Secanell, J. Power Sources **206**, 153–160 (2012)
25. J.D. Ramshaw, J. Non-Equilib. Thermodyn. **15**, 295–300 (1990)
26. Z. Yu, R.N. Carter, J. Power Sources **195**, 1079–1084 (2010)
27. W. He, B. Wang, H. Zhao, Y. Jiao, J. Power Sources **196**, 9985 (2011)
28. H. Matsumoto, I. Nomura, S. Okada, T. Ishihara, Solid State Ionics **179**, 1486–1489 (2008)
29. C. Chan, N. Zamel, X. Li, J. Shen, Electrochim. Acta **65**, 13–21 (2012)

Chapter 3
Diffusivity Measurement Techniques

One of the major activities in the field of mass transport of solid oxide fuel cells (SOFCs) is the development of facile diffusivity measurement techniques. In this chapter, techniques for diffusivity measurement of porous electrodes are overviewed, with an emphasis on the authors' contributions in this area. The correlations of the diffusivity with concentration polarizations in SOFCs also are illustrated thoroughly in this chapter.

3.1 Diffusivity Measurement in Porous Media

Recently, different techniques have been developed to measure gas diffusivity in fuel cells. Although each measurement protocol was designed for a specific type of fuel cell, the investigated devices could be employed for the gas diffusivity measurements in other types of fuel cells due to their similarity in fuel types and architectural designs. Therefore, an overview of gas diffusivity measurement techniques in other fuel cells, such as PEMFCs, is beneficial for the development of gas diffusivity measurements and of energy loss evaluation in SOFCs. For PEMFCs, three types of measurement devices have been investigated recently. Figure 3.1 shows a self-heated oxygen sensor device for the gas diffusivity measurement in PEMFCs [1].

N_2 and air channels were used in the device to create oxygen pressure gradients and oxygen diffusion through the electrode between the two channels. For a given value of electrode thickness (δ), diffusive length scale, width (m) of the sealing gasket used in the device, gas flow rate, and oxygen pressure at the exhaust port, the gas diffusivity can be calculated through Eq. 3.1,

$$D_{O_2}^{\text{eff}} = \frac{mQC_{x=L}}{2\delta L C_{\text{air}}} \tag{3.1}$$

where $C_{x=L}$ is molar oxygen concentration at the exhaust of N_2 channel and C_{air} is molar oxygen concentration at the exhaust of N_2 channel in the air

© The Author(s) 2014

W. He et al., *Gas Transport in Solid Oxide Fuel Cells*, SpringerBriefs in Energy,
DOI 10.1007/978-3-319-09737-4_3

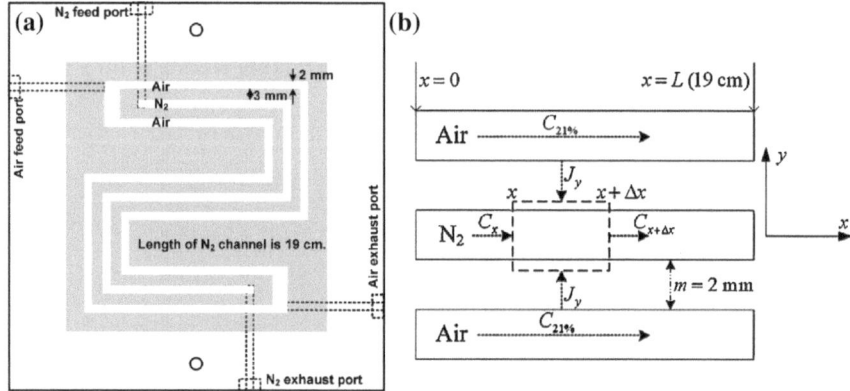

Fig. 3.1 **a** Plan view of a self-heated oxygen sensor device and **b** schematic of oxygen mass balance through N_2 channel. Reprinted from Ref. [1]. Copyright (2010), with permission from Elsevier

Fig. 3.2 Schematic of the modified Loschmidt cell for the measurement of effective diffusion coefficient. *1* gas inlet 1; *4* gas inlet 4; *2, 3,* and *5* outlets; *6* sliding gate valve; *6a* open position of the valve; *6b* closed position of the valve; *7* oxygen probe; *8* and *9* thermocouples; *10* sample holder. Reprinted from Ref. [2]. Copyright (2012), with permission from Elsevier

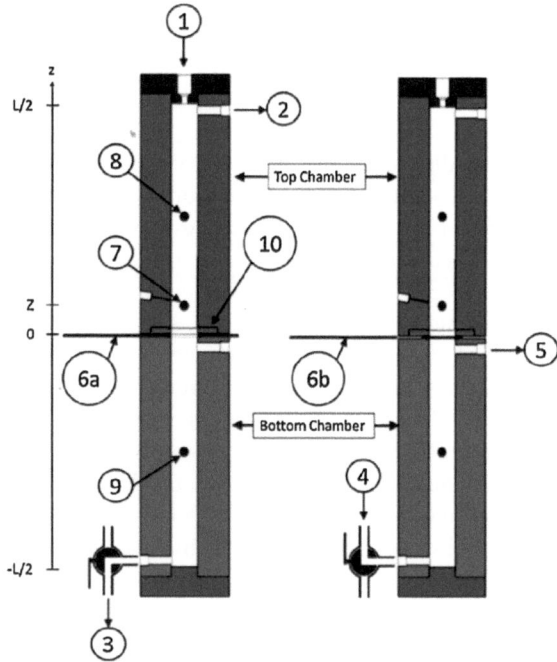

channel. Another type of gas diffusivity measurement device based on a modified Loschmidt cell is shown in Fig. 3.2 [2].

For a given value of oxygen pressure in the device measured with an oxygen probe, the effective diffusion coefficient of a porous layer is obtained via Eq. 3.2,

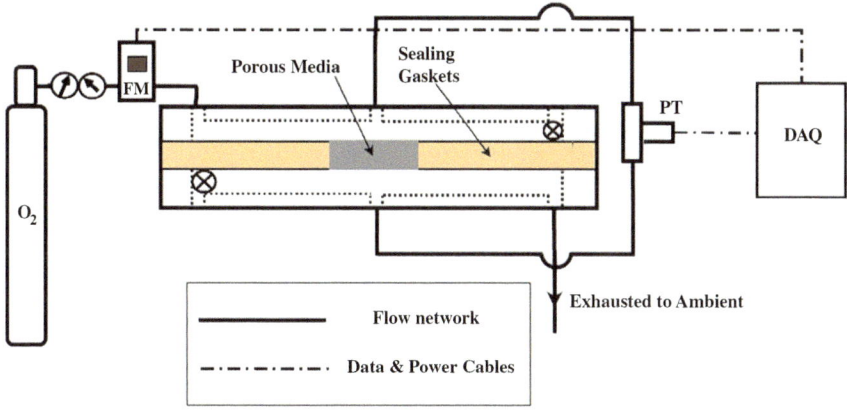

Fig. 3.3 Schematic of the permeability measurement experimental setup, *FM* mass flow controller; *PT* differential pressure transducer; *DAQ* data acquisition card. Reprinted from Ref. [5]. Copyright (2012), with permission from Elsevier

$$D_i^{\text{eff},1} = \frac{L_2 - L_0}{H/D_i^{\text{eq}} - (H - L_2 - L_0)/D_i^{\text{bulk}}} \tag{3.2}$$

where D_i^{eq} is the equivalent diffusivity of species i and is obtained via Eq. 3.3 [3].

$$C_i = \frac{C_i^b}{2} \text{erfc}\left(\frac{H}{2\sqrt{\Delta t D_i^{\text{eq}}}}\right) \tag{3.3}$$

In Eq. 3.16, D_i^{bulk} is bulk diffusivity of species i and can be calculated through Eq. 3.4 [4].

$$\ln D_i^{\text{bulk}} = 1.724 \ln T + \ln(1.13 \times 10^{-5}) - \ln p \tag{3.4}$$

Until now, only room temperature diffusivity has been measured using this device. However, owing to the robustness of its components, high-temperature measurements are also possible with this device.

Another device based on a modified diffusion bridge setup was recently employed to measure the Knudsen diffusivity and absolute permeability in PEMFC electrodes [4–6]. The schematic of the device is shown in Fig. 3.3. A porous electrode is sandwiched between two gas channels. A gas pressure transducer was employed to trace the pressure gradient of gas diffusion through the porous electrode. A BFM model was employed to calculate the gas diffusivity, as well as the correlation between absolute permeability and gas diffusivity [5]. With some modifications, such as replacing the components with high-temperature ceramics, different devices for the measurement of gas diffusivity in PEMFCs could be employed for such measurements in SOFCs. A review on recently developed electrochemical devices for gas diffusivity measurements in SOFCs will be presented next.

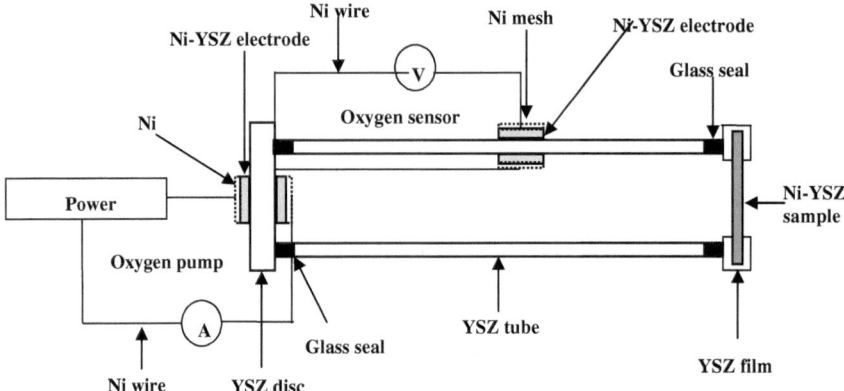

Fig. 3.4 Single-sensor electrochemical device for anode gas diffusivity measurement in SOFCs [7]. Reprinted from Ref. [7]. Copyright (2010), with permission from Elsevier

3.2 Advanced Diffusivity Measurement Techniques in Solid Oxide Fuel Cells

To facilitate anode gas diffusivity measurements in SOFCs, oxygen-sensor- and oxygen-pump-based electrochemical devices, as shown in Fig. 3.4, developed [7]. To make a device, a YSZ tube was employed, with a YSZ disk attached to one end and an anode/cathode sample attached to the other. An oxygen pump was employed to provide current across the YSZ disk. A voltage meter was used as an oxygen sensor to measure the oxygen pressure gradient across the YSZ tube. YSZ disk–YSZ tube and electrode–YSZ tube connections were accomplished using glass paste, and the connection between wiring and YSZ was achieved with Ni-YSZ paste and glue. The device was employed to measure the binary effective H_2/H_2O gas diffusivity in SOFCs at operating conditions. During the measurement, H_2O/H_2 was introduced into an airtight tube furnace where the measurement setup was placed. Upon supply of a current i via the oxygen pump, a flux of H_2/H_2O was induced through the anode sample at the other end of the YSZ tube. The mathematical correlation between the current supply and the induced flux is expressed by Eq. 3.5.

$$J_{H_2} = -J_{H_2O} = \frac{i}{4F} \qquad (3.5)$$

Flux is a function of the diffusivity and the H_2 pressure gradient across the anode sample, as shown in Eq. 3.6,

$$J_{H_2} = \frac{i}{4F} = -D_{H_2-H_2O}^{eff} \nabla n_{H_2} = -\frac{D_{H_2-H_2O}^{eff}}{RT} \frac{dp_{H_2}}{dx} \qquad (3.6)$$

where $D_{H_2-H_2O}^{eff}$ is the effective binary diffusivity of H_2 and H_2O, ∇n_{H_2} is the concentration difference within dx, and dp_{H_2} is the pressure gradient of H_2 within dx.

Integration over H_2 pressure from one side of the anode sample to the other side of the anode sample leads to Eq. 3.7,

$$p^i_{H_2} = p^o_{H_2} - \frac{RTl_a}{4FD^{eff}_{H_2-H_2O}} i \qquad (3.7)$$

where $p^i_{H_2}$ is the H_2 pressure in the YSZ tube, $p^o_{H_2}$ is the H_2 pressure out of YSZ tube, and l_a is the anode thickness.

Equation 3.7 shows that the H_2 pressure in the YSZ tube is a function of i. At a certain temperature, the reaction equilibrium constant for H_2O, H_2, and O_2 is fixed. With the pressure difference across the tube measured via a voltage meter, the effective binary H_2/H_2O diffusivity can be expressed by Eq. 3.7. With the measured diffusivity, the anode concentration polarization loss, and the anode limiting current can be calculated. The setup works highly efficiently; <5 min was sufficient for the sensor to read a stable voltage value [47].

Similar to the anode diffusivity measurement, a cathode diffusivity measurement was conducted by Zhao et al., using an electrochemical cell [8]. Using the device, the effective binary O_2-N_2 diffusivity was measured with O_2-N_2 flow at operating conditions. In this study, the correlation among concentration and a few other important parameters, such as cathode thickness, and porosity, was studied. For a specific porosity, CP increased with increasing cathode thickness, whereas at a fixed temperature and cathode thickness, CP decreased with increasing cathode porosity. Therefore, the authors proposed using thin porous cathodes to improve O_2 diffusivity and, in turn, enchance the efficiency of SOFCs. The effective binary O_2-N_2 diffusivity was found to follow a power law, $D = aT^{3/2}$, where a, a constant, increased with decreasing cathode porosity. Gas diffusivity within high-porosity cathodes in SOFCs is less sensitive to the variation of operating temperature compared to low-porosity cathodes. Therefore, using highly porous cathodes allows the operating temperature to be varied to meet practical requirements without increasing the concentration polarization in the operation of an SOFC.

Direct measurement of the gas diffusivity in SOFCs via a single-sensor electrochemical cell greatly facilitates the efficient evaluation of an electrode before intact fuel cells are fabricated. However, a single-sensor electrochemical device still needs improvement to be considered for diffusivity measurements. For instance, only one sample can be tested at a time, and additional heating/cooling and gas switches are needed to measure additional electrodes. As noticed from Fig. 3.4, only one electrode sample is attached to the YSZ disk, leaving the other side of the YSZ disk unutilized. The authors of that study proposed that two YSZ tubes be attached to both sides for the simultaneous testing of two electrodes [9]. Figure 3.5 shows a schematic of the device. The device doubled the efficiency of diffusivity measurements in SOFCs and facilitated the evaluation of the electrode thickness more efficiently compared to the single-sensor electrochemical device.

If we consider a single-sensor electrochemical cell to be zero-dimensional, and a double-sensor electrochemical cell to be one-dimensional, then two-dimensional and three-dimensional electrochemical cells should show higher efficiency for electrode

Fig. 3.5 Double-sensor
electrochemical device for
anode/cathode gas diffusivity
measurements in SOFCs.
Reprinted from Ref. [9].
Copyright (2011), with
permission from Elsevier

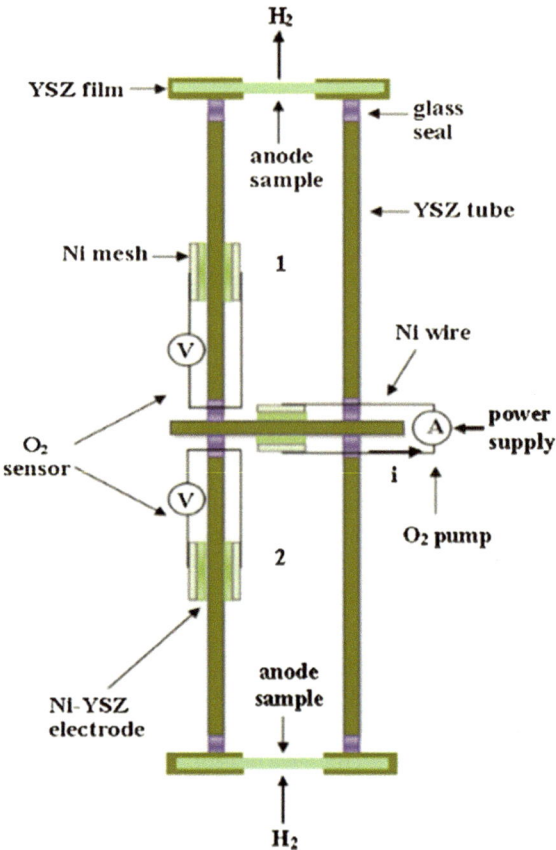

diffusivity measurements in SOFCs compared to the single-/double-sensor devices. Further, increasing the diffusivity measurement efficiency allows one to more extensively investigate the CP and the limiting current density with respect to other important parameters associated with fuel cells, such as the applied current density, the thickness of nanostructured fuel cells electrodes, and the electrode porosity. A multi-sensor electrochemical device was designed by using a three-dimensional electrolyte cube, as shown in Fig. 3.6 [10]. For YSZ-based SOFCs, six YSZ tubes can be attached to the six sides of the cube, enabling simultaneous testing of six electrodes.

To enhance the efficiency of diffusivity measurement further and to evaluate the correlation between CP and the important electrode parameters on the nanoscale, an electrochemical cell with a large electrolyte plate was proposed to which many electrolyte tubes can be attached simultaneously [11]. The device for the gas diffusivity measurement in SOFCs is shown in Fig. 3.7. Similar to the YSZ cube used in a previously reported three-dimensional device, the two-dimensional electrolyte plate allows multiple tubes and electrodes to be loaded and measured at the same time [11]. One can conveniently design the loading capacity of the device by adjusting the surface area of the plate. Further, the plate-based device makes the

Fig. 3.6 **a** A YSZ tube. **b** A Schematic of the Multi-sensor electrochemical device for anode/cathode gas diffusivity measurement in SOFCs. The directions of the H_2 fluxes and the current provided by the oxygen pump are indicated by *arrows*. Reprinted from Ref. [10]. Copyright (2012), with permission from Wiley-VCH Vertag GmbH

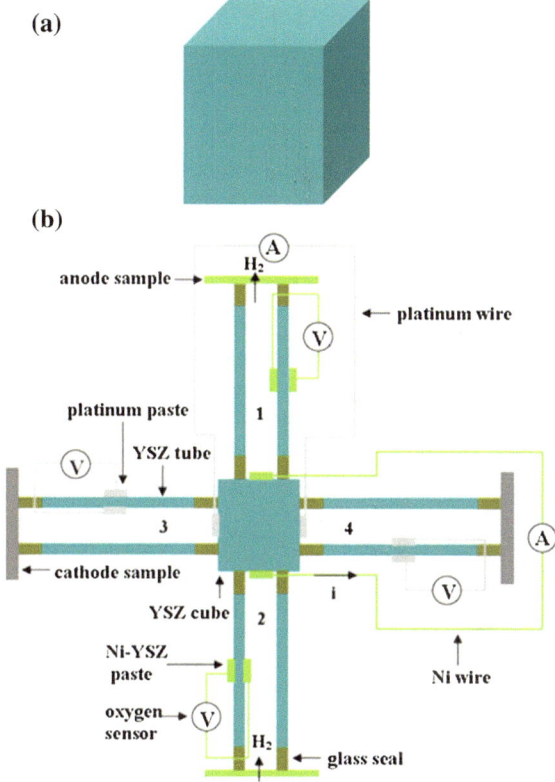

Fig. 3.7 A schematic of an electrochemical device with a 2D electrolyte plate for the measurement of the effective binary diffusivity in *solid* oxide fuel cells. Reprinted from Ref. [11]. Copyright (2012), with permission from Elsevier

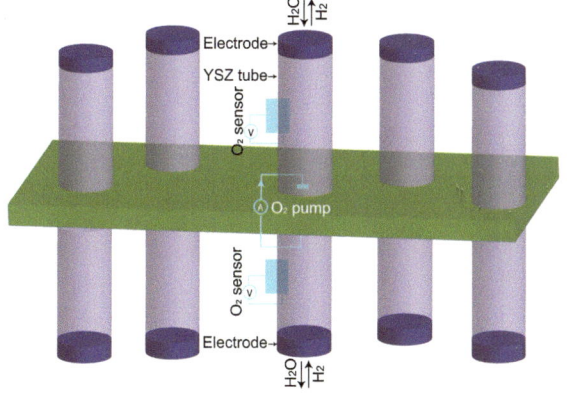

manipulation of the wiring for the multiple oxygen pumps and sensors more convenient compared to the reported three-dimensional device.

Recent research on the experimental and theoretical 3D microstructure reconstruction of fuel cell electrodes has confirmed the existence of porosity, connectivity, and tortuosity anisotropies in the three directions of these electrodes [12–14]. The experimental realization of high-performance electrodes and theoretical insight into 3D diffusivity have facilitated both the development of low-energy fuel cells and a fundamental understanding of gas diffusion in porous electrodes. An experimental technique for single-step measurements of the 3D diffusivity in fuel cells is highly desirable allowing for the efficient determination of the optimum diffusion direction among the three directions of an electrode. Efficient assessment of the optimum operation diffusion direction is the basis of efficiently selecting the electrode direction after a raw porous electrode is prepared via deposition, die-pressing, and other methods. In addition, 3D diffusivity measurements can directly lead to both the evaluation of the concentration polarization in all three directions of an electrode and the quantitative investigation of the correlation among concentration polarization energy loss and the 3D porosity, tortuosity, and other electrode parameters [15–21]. A schematic of the electrochemical device to realize a 3D diffusivity measurement is shown in Fig. 3.8a [22]. For an anode gas diffusivity measurement in a SOFC, a cube made of an anode material to be tested is employed. Two YSZ tubes are attached to the two parallel sides of the electrode

Fig. 3.8 **a** An oxygen-sensor-based electrochemical device for 3D diffusivity measurement in fuel cells, **b** an oxygen-pump-based electrochemical device, and **c** an improved oxygen-sensor-based electrochemical cell for 3D diffusivity measurement in fuel cells. Reprinted from Ref. [22]. Copyright (2013), with permission from Elsevier

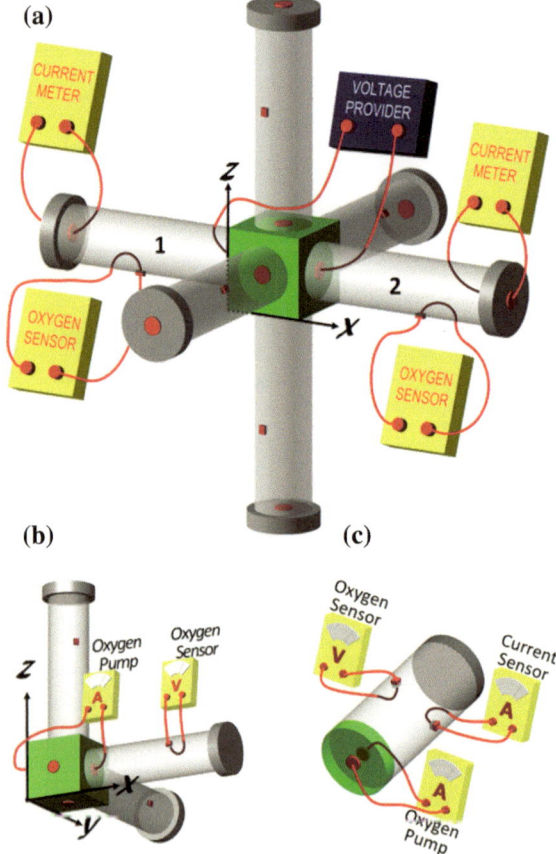

cube, and a voltage provider is placed on these two sides along x-axis. On the other end of each YSZ tube, a YSZ disk is attached; a current meter is placed across the YSZ disk. The cube–tube connection is produced using glass paste, and the current meter, oxygen sensor, and voltage provider are attached to the disk and the cube using a Ni-YSZ electrode paste mixed with sealing glue. In a similar fashion, four additional YSZ tubes can be attached to the other four cube sides. With such a device, electrode gas diffusivity in all three electrode directions can be measured simultaneously. The direction and the magnitude of the summed 3D diffusivity are readily determined from the summing principle of vertical vectors if it is assumed that the gas pressure gradients in the three electrode directions are the same, as shown in Eq. 3.8,

$$D_s = \sqrt{D_x^2 + D_y^2 + D_z^2} \tag{3.8}$$

where D_x, D_y, and D_z are the diffusivities along three electrode directions, with $D_y = aD_x$ and $D_z = bD_x$, respectively [22]. With the measured 3D diffusivity, 3D CP, 3D porosity, and 3D tortuosity can be evaluated. Thus, the optimum operation direction of a raw electrode sample can be determined before the assembly and operation of the electrode in a fuel cell system.

3.3   The Role of Advanced Diffusivity Measurement Techniques in Exploring Highly Efficient Solid Oxide Fuel Cell Electrodes

3.3.1   Correlations between the Diffusivity and Concentration Polarization

In the traditional method of evaluating energy losses in SOFCs, CP, OL, and AL are obtained by fitting the multiple voltage–current (V–I) data measured under operating conditions according to Eq. 3.9,

$$
\begin{aligned}
E(i) = E_{ocp} - iR_i &- \frac{2RT}{F} \ln\left\{\frac{1}{2}\left[\left(\frac{i}{i_o}\right) + \sqrt{\left(\frac{i}{i_o}\right)^2 + 4}\right]\right\} \\
&+ \frac{RT}{2F} \ln\left(1 - \frac{i}{i_a}\right) - \frac{RT}{2F} \ln\left(1 + \frac{p_{H_2}^o i}{p_{H_2O}^o i_a}\right) + \frac{RT}{4F} \ln\left(1 - \frac{i}{i_c}\right)
\end{aligned}
\tag{3.9}
$$

where E_{ocp} is the open-circuit potential, R_i is the Ohmic resistance, F is the Faraday's constant, p is the gas pressure, i_o is the exchange current density, i_a is the anode limiting current density, and i_c is the cathode limiting current density [23]. While this method is mathematically feasible, the accuracy of fitting five parameters based on a limited number of measurements is debatable. The correlation between the concentration polarization and the limiting current density is shown in Eqs. 3.10 and 3.11 [24],

$$\eta_a = \frac{RT}{2F} \ln\left(1 - \frac{i}{i_a}\right) + \frac{RT}{2F} \ln\left(1 - \frac{p_{H_2}^o i}{p_{H_2O}^o i_a}\right) \tag{3.10}$$

$$\eta_c = \frac{RT}{4F} \ln\left(1 - \frac{i}{i_c}\right) \tag{3.11}$$

The limiting current density is a function of the diffusion coefficient associated with anode/cathode gas transport, i.e., diffusivity, as shown in Eqs. 3.12 and 3.13 [24],

$$i_a \approx \frac{2Fp_{H_2}^o D_{H_2-H_2O}^{\mathrm{eff}}}{RTl_a} \tag{3.12}$$

$$i_c \approx \frac{4Fp_{O_2}^o D_{O_2-N_2}^{\mathrm{eff}}}{RTl_c}\left(\frac{p_t}{p_t - p_{O_2}^o}\right) \tag{3.13}$$

where F is the Faraday's constant, $D_{H_2-H_2O}^{\mathrm{eff}}(D_{O_2-N_2}^{\mathrm{eff}})$ is the effective binary diffusivity of $H_2/H_2O(O_2/N_2)$ gas, and $l_a(l_c)$ is the anode(cathode) thickness. With known diffusivity, the anode/cathode concentration polarization can be calculated directly; direct measurement of gas diffusivity in the anode/cathode of a fuel cell facilitates the reliable evaluation of the concentration polarization of a fuel cell.

Figure 3.9 depicts the plots of anode/cathode concentration polarization as a function of the diffusivity for different fuel cell operating temperatures and different electrode thicknesses. For both anodes and cathodes, the electrode CP decreases with increasing electrode diffusivity. At high operating temperatures, the CP appears to decrease with increasing diffusivity more rapidly. This phenomenon suggests that the CP at a high operating temperature is more sensitive to the change in electrode diffusivity compared to the CP at a relatively low operating temperature. Such a quantitative analysis of the CP versus diffusivity is significantly important for the rational design of the electrode thickness and the operating temperature for a fuel cell system.

To consider the relationship among the polarization concentration calculated using the measured diffusivity with the operating current density and the diffusivity,

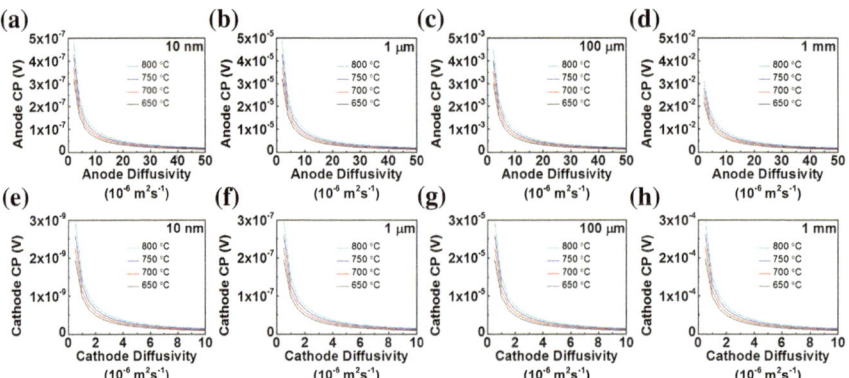

Fig. 3.9 The plots of electrode concentration polarization versus electrode diffusivity with different operating temperatures. In the CP calculations, $p_{O_2}^o = 0.21$ atm, $p_{H_2}^o = 0.03$ atm, and $p_t = 1$ atm. Reprinted from Ref. [22]. Copyright (2013), with permission from Elsevier

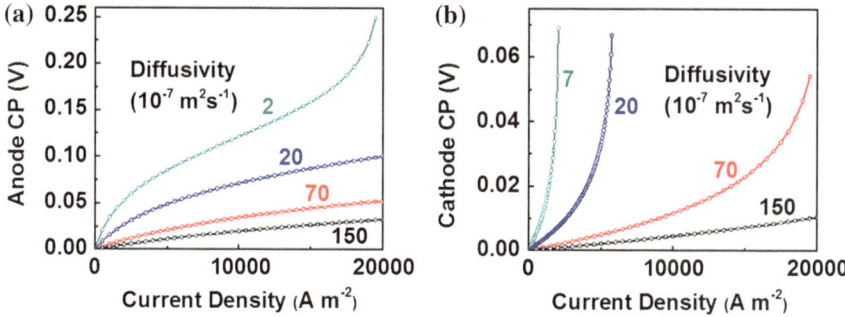

Fig. 3.10 The plots of **a** anode and **b** cathode concentration polarization versus current density with different diffusivities. Reprinted from Ref. [22]. Copyright (2013), with permission from Elsevier

CP versus applied current density with different gas diffusivities was plotted, as shown in Fig. 3.10. In Fig. 3.10a, the CP for a 2-mm anode increases with increasing applied current density for all the anode diffusivities. For $2 \times 10^{-7}, 7 \times 10^{-7}$, and 15×10^{-7} m^2 s^{-1} anode diffusivities, the CP increases rapidly with applied current density; the increase appears to be more gradual as the applied current density reaches a certain value. If we let the transition value as critical applied current density (i_{cr}), the i_{cr} for anode diffusivities $2 \times 10^{-7}, 7 \times 10^{-7}$, and 15×10^{-7} m^2 s^{-1} is 3,000, 7,200, and 10,000 A m^{-2}, respectively. That the critical current density increases with increasing anode diffusivity suggests that the anode CP is less sensitive to an increase in the applied current density for anodes with larger diffusivities compared to anodes with smaller diffusivities. For anodes with a diffusivity of 2×10^{-7} m^2 s^{-1}, in addition to much larger CPs, the anode CP versus i plot for this anode diffusivity shows an inflection point at 18,000 A m^{-2}, above which anode CP reaches an infinite value abruptly. Therefore, working current densities in the range of 5,000–18,000 A m^{-2} are proposed for the anode with diffusivity around 2×10^{-7} m^2 s^{-1} to ensure a relative low concentration polarization energy loss when such low anode gas diffusivity is employed in a fuel cell system. For cathodes, except for those with a very large diffusivity (150×10^{-7} m^2 s^{-1}), the CP reaches infinite values rapidly with medium or small cathode diffusivities, as shown in Fig. 3.10b. Thus, compared to anodes, the cathode CP tends to be more dominant in practical ranges of operating current densities.

3.3.2 Correlations Between Concentration Polarization and Structures of Anodes/Cathodes

With the measured anode/cathode diffusivities, the correlation between the anode/cathode CP and electrode properties, such as thickness, porosity, and tortuosity can be built. The plotted CP–thickness curves displayed in Fig. 3.11 for 650, 700, 750, and 800 °C show the crossing features for both cathodes and anodes. The findings were surprising because larger T typically showed smaller CP for SOFCs, as

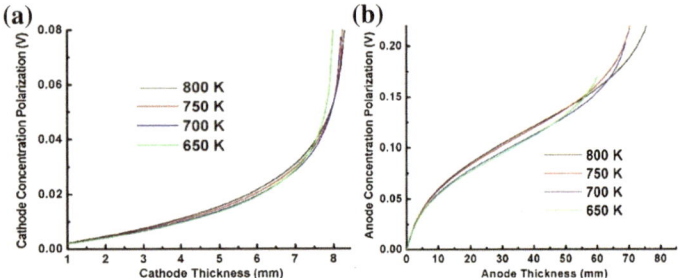

Fig. 3.11 **a** Correlation between CP and cathode thickness and **b** correlation between CP and anode thickness. Reprinted from Ref. [9]. Copyright (2011), with permission from Elsevier

previously observed. For example, with a fixed anode thickness of 0.5 mm, and an applied current density of 0.05 A cm^{-2}, among the four temperatures, the anode CP at 750 °C was the largest, followed by the CPs at 800, 700, and 650 °C, respectively. For anode thicknesses above 10 mm, the anode CP increased with increasing operating temperature. A similar phenomenon was observed in the cathode CP versus cathode thickness plots. As shown in Fig. 3.11, the anode CP appears to increase very rapidly with increasing anode thickness, while the cathode CP increases very gradually with increasing cathode thickness. Such findings are based on efficient diffusivity measurements with the double-sensor electrochemical device, and the findings provide an efficient platform for one to pre-evaluate anode/cathode thickness before the electrode is fabricated and assembled.

The effective diffusivity is a function of electrode porosity and tortuosity, as shown in Eqs. 3.14 and 3.15.

$$D_{H_2-H_2O}^{eff} = \frac{\phi_a}{\tau_a} D_{H_2-H_2O} \tag{3.14}$$

$$D_{O_2-N_2}^{eff} = \frac{\phi_c}{\tau_c} D_{O_2-N_2} \tag{3.15}$$

For a fixed tortuosity, operating temperature, and thickness, the diffusivity is linearly dependent on electrode porosity. Thus, a direct measurement of the gas diffusivity in an SOFC electrode largely facilitates the efficient evaluation of the correlation between concentration polarization and electrode porosity. The correlation between CP and porosity for fuel cells with nanostructured electrodes is shown in Fig. 3.12 [25]. As seen in Fig. 3.12, CP decreases with increasing the anode porosity for 10-, 100-, 500-, and 1,000-nm-thick anode samples at 650, 700, 750, and 800 °C. The monotonous decrease of CP with an increase in electrode porosity confirms that high porosity is favorable to improve the energy conversion efficiency of SOFCs.

For most porous fuel cell electrodes, the value of the tortuosity ranges from 2 to 25. The fuel cell performance can be improved by tuning the tortuosity values of the fuel cell electrodes within or beyond the range [8]. The relationship between the effective binary gas diffusivity and tortuosity is then studied. Figure 3.13 shows the electrode diffusivity versus the tortuosity plots for different electrode porosities. For both anodes

Fig. 3.12 Correlation between CP and anode porosity for **a** anode thickness: 10 nm, **b** anode thickness: 100 nm, **c** anode thickness: 500 nm, and **d** anode thickness: 1,000 nm. Reprinted from Ref. [11]. Copyright (2012), with permission from Elsevier

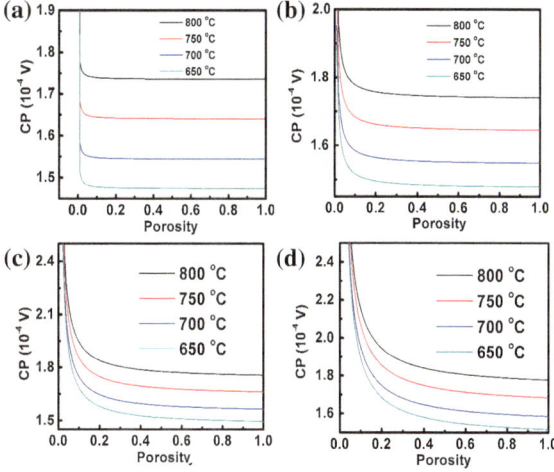

Fig. 3.13 The plots of anode (**a**) cathode (**b**) effective binary diffusivity versus tortuosity with different electrode porosities and a fixed electrode thickness of 1 mm. Reprinted from Ref. [22]. Copyright (2013), with permission from Elsevier

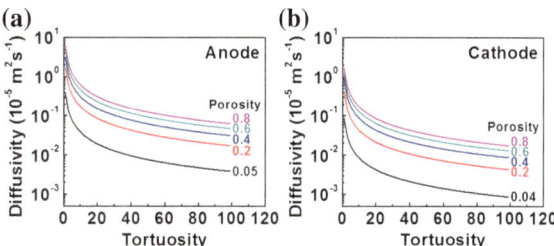

and cathodes, diffusivity increases with decreasing tortuosity. Below a critical tortuosity, the diffusivity increases rapidly and then reaches an infinite value. The critical tortuosity increases with the increase of electrode porosity; thus, to increase gas diffusivity, it is more efficient to decrease the tortuosity of an electrode with larger porosity than to decrease the tortuosity of an electrode with smaller porosity. Comparing with the diffusivity versus tortuosity plots for anodes, the diffusivity versus tortuosity plots for cathodes appear to be more porosity dependent, which indicates that increasing cathode porosity is more efficient to increase the gas diffusivity and to reduce the concentration polarization of a fuel cell system than increasing anode porosity.

3.4 Quantity Analysis of Measurement Error of the Diffusivity and Concentration Polarization

3.4.1 Current Error

In the just discussed diffusivity measurements of the electrochemical devices, the electronic conduction contribution of electrolyte materials was neglected, and an error in the subsequent evaluations of the limiting current density and concentration polarization was consequently induced. The evaluation error of the effective

binary $H_2/H_2O(O_2/N_2)$ diffusivity induced by current measurement uncertainty is
expressed in Eqs. 3.16 and 3.17 [26],

$$\Delta D_{H_2-H_2O}^{eff} = \frac{RTl_a}{4F(p_{H_2}^o - p_{H_2}^i)} \Delta i \tag{3.16}$$

$$\Delta D_{O_2-N_2}^{eff} = \frac{RTl_c}{8F(p_{O_2}^i - p_{O_2}^o)} \Delta i \tag{3.17}$$

where $p_{H_2}^i(p_{O_2}^i)$ is the H_2 (O_2) pressure in the YSZ tube and $p_{O_2}^o$ is the pressure of
cathode gas. $p_{H_2}^i$ can be obtained via the voltage measurement using the oxygen sensor and with the reaction equilibrium constant of $H_2O/H_2/O_2$ system at a specific
temperature [9]. ΔD is linearly dependent on Δi as a certain current is provided via
the oxygen pump, as shown in Eqs. 3.16 and 3.17. With $p_{H_2}^i$ set as a temperature-
dependent parameter, ΔD versus $\Delta i/i$ dependences at different temperatures can be
readily plotted. As shown in Fig. 3.14, for both the anode (Fig. 3.14a) and the cathode
(Fig. 3.14b), ΔD versus $\Delta i/i$ plots show obvious slopes; a 10 % uncertainty in the current measurement induces a 0.006 cm^2 s^{-1} (0.007 cm^2 s^{-1}) error in anode(cathode)
gas diffusivity measurement at 700 °C, indicating that the accuracy of anode/cathode
diffusivity measurement is highly sensitive to the uncertainty of current measurement.
One can notice that with a fixed current uncertainty, ΔD increases with increasing
temperature for both anodes and cathodes.

The limiting current density is an important parameter for energy conversion systems and correlates with the current utilization and the polarization loss of fuel cells

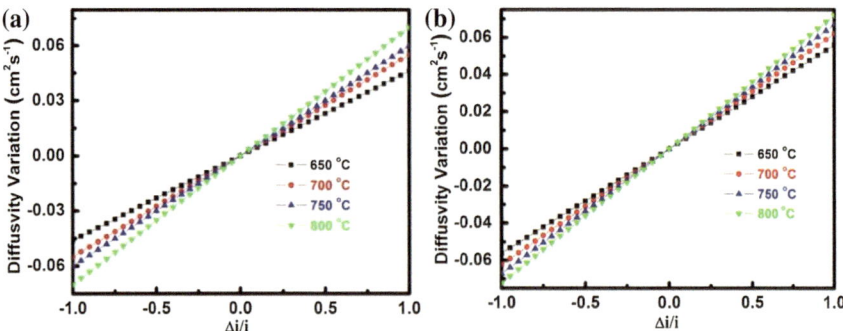

Fig. 3.14 Anode (**a**) and cathode (**b**) diffusivity measurement errors as a function of current
measurement uncertainty ($\Delta i/i$) induced by ignoring electronic conduction contribution of electrolyte with an applied current density of 500 A m^{-2}. Anode thickness is 0.75 mm, and cathode
thickness is 0.2 mm. Anode hydrogen effective binary diffusivities are 0.07 cm^2 s^{-1} for 800 °C,
0.06 cm^2 s^{-1} for 750 °C, 0.055 cm^2 s^{-1} for 700 °C, and 0.046 cm^2 s^{-1} for 650 °C, respectively.
Cathode oxygen effective binary diffusivities are 0.072 cm^2 s^{-1} for 800 °C, 0.067 cm^2 s^{-1} for
750 °C, 0.062 cm^2 s^{-1} for 700 °C, and 0.056 cm^2 s^{-1} for 650 °C, respectively. Reprinted from
Ref. [26]. Copyright (2010), with permission from Elsevier

[27–30]. Efficient diffusivity measurements directly lead to the evaluation on the limiting current densities of a fuel cell system. Anode limiting current density i_a and the cathode limiting current density i_c exhibit a linear dependence on diffusivity; thus, $\Delta i_a (\Delta i_c)$ also shows a linear dependence on Δi, as shown in Eqs. 3.18 and 3.19, [26]

$$\Delta i_a \approx \frac{4 F p_{H_2}^o \Delta D_{H_2 - H_2 O}^{eff}}{R T l_a} \tag{3.18}$$

$$\Delta i_c \approx \frac{8 F p_{O_2}^o \Delta D_{O_2 - N_2}^{eff}}{R T l_c} \left(\frac{p_t}{p_t - p_{O_2}^o} \right) \tag{3.19}$$

where P_t (~1 atm) is the total gas pressure in the measurement system. Based on Eqs. 3.18 and 3.19, $\Delta i_a (\Delta i_c)$ can be plotted as a function of Δi. As shown in Fig. 3.15, similar to the $\Delta D_{H_2 - H_2 O}^{eff} (\Delta D_{O_2 - N_2}^{eff})$ versus $\Delta i / i$ plots, $\Delta i_a (\Delta i_c)$ versus $\Delta i / i$ plots also show notable slopes; for anodes, the slope for 800 °C is 1646.3 A m^{-2}, followed by 1505.2 A m^{-2} for 750 °C, 1478.3 A m^{-2} for 700 °C, and 1331.5 A m^{-2} for 650 °C, respectively. Compared with the ΔD versus $\Delta i / i$ plots, $\Delta i_a (\Delta i_c)$ shows a weaker temperature dependence. For instance, Δi_c does not show an obvious temperature dependence, as confirmed by the similar slopes that the Δi_c versus $\Delta i / i$ plots at different temperatures have. The slope of the Δi_c versus $\Delta i / i$ plot only varies by 4.4 % as the operating temperature changes from 650 to 800 °C. Comparing anodes and cathodes, while Δi_a versus $\Delta i / i$ plots at different temperatures are still distinguishable, the Δi_c versus $\Delta i / i$ plots at different temperatures almost overlap. This implies that lowering the operating temperature cannot effectively reduce the measurement error of the cathode limiting current density, induced by the current uncertainty in the cathode gas diffusivity measurement.

To improve the electrochemical performance and mechanical strength of fuel cells, nanostructured electrodes and electrolytes have been proposed and investigated [31–41]. In particular, thin electrodes have been proven to possess many advantages over bulk electrodes, such as high gas diffusion rate, larger limiting current density, and low polarization loss, among others [32, 33]. Therefore, an analytical evaluation of the diffusivity measurement error in nanostructured fuel cells is of practical significance. Figure 3.16c and d shows $Log(\Delta i_a)(Log(\Delta i_c))$ versus $\Delta i / i$ plots for the anode (cathode) of a fuel cell. With a fixed current measurement uncertainty, the plots for the four different electrode thicknesses follow similar trends, and the evaluation error of limiting current density increases with reducing the anode (cathode) thickness. With the same Δi, $Log(\Delta i_a)(Log(\Delta i_c))$ for 75-nm anode (20-nm cathode) is the largest among the four thicknesses, followed by those for 7.5-μm anode (2-μm cathode), 750-μm anode (200-μm cathode), and 7.5-mm anode (2-mm cathode), respectively.

The reliability of the gas diffusivity measurement directly impacts the accuracy of the polarization loss pre-evaluation. The correlation between the evaluation error of the anode(cathode) concentration polarization loss and the current measurement uncertainty is described in Eqs. 3.20 and 3.21 [26],

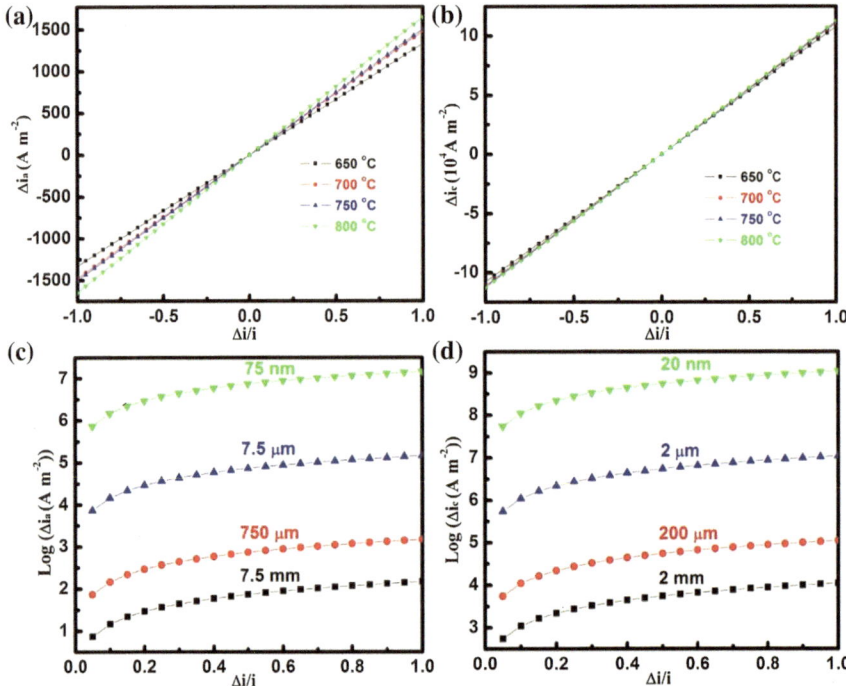

Fig. 3.15 Anode (**a**) and cathode (**b**) limiting current density evaluation errors as a function of current measurement uncertainty ($\Delta i/i$) induced by ignoring electronic conduction contribution of electrolyte with an applied current density of 500 A m^{-2}. Anode thickness is 0.75 mm, and cathode thickness is 0.2 mm. Anode limiting current densities are 1.65×10^3 A m^{-2} for 800 °C, 1.51×10^3 A m^{-2} for 750 °C, 1.48×10^3 A m^{-2} for 700 °C, and 1.33×10^3 A m^{-2} for 650 °C, respectively. Cathode limiting current densities are 11.25×10^4 A m^{-2} for 800 °C, 11.17×10^4 A m^{-2} for 750 °C, 11.07×10^4 A m^{-2} for 700 °C, and 10.77×10^4 A m^{-2} for 650 °C, respectively. Log(Δi_a)(Log(Δi_c)) versus $\Delta i/i$ plots for different anode (**c**) (cathode (**d**)) thicknesses with an operating temperature of 700 °C and an applied current density of 500 A m^{-2}. Reprinted from Ref. [26]. Copyright (2010), with permission from Elsevier

$$\Delta \eta_a = -\frac{RT}{2F} \ln \left(1 - \frac{i}{\Delta i_a} \right) + \frac{RT}{2F} \ln \left(1 + \frac{p^o_{H_2} i}{p^o_{H_2O} \Delta i_a} \right) \qquad (3.20)$$

$$\Delta \eta_c = -\frac{RT}{4F} \ln \left(1 - \frac{i}{\Delta i_c} \right) \qquad (3.21)$$

where $\Delta \eta_a (\Delta \eta_c)$ is the anode (cathode) concentration polarization evaluation error induced by the current value uncertainty in the anode (cathode) gas diffusivity measurement via previous electrochemical devices. Figure 3.16a, b shows the ΔCP versus $\Delta i_a/i$ plots for anodes and cathodes at different temperatures. The plots for the cathodes resemble the Δi_a (Δi_c) versus $\Delta i/i$ plots, but different from the $\Delta i_a(\Delta i_c)$ versus $\Delta i/i$ plots. For the four temperatures, the slopes of ΔCP

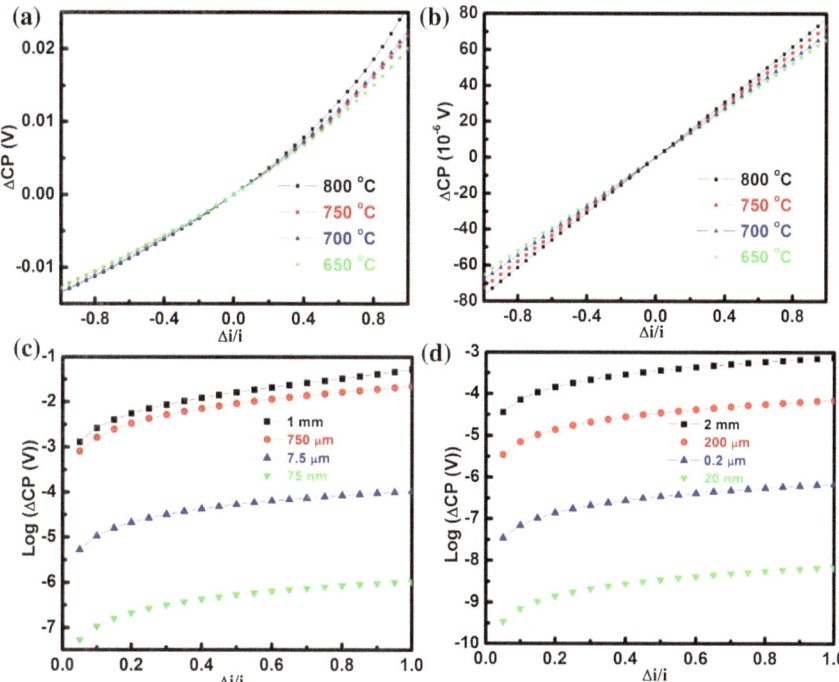

Fig. 3.16 Anode (**a**) and cathode (**b**) concentration polarization evaluation errors as a function of current measurement uncertainty ($\Delta i/i$) induced by ignoring electronic conduction contribution of electrolyte. Anode thickness is 0.75 mm, and cathode thickness is 0.2 mm. Anode concentration polarizations are 0.0135 V for 800 °C, 0.0128 V for 750 °C, 0.0134 V for 700 °C, and 0.0128 V for 650 °C, respectively. Cathode concentration polarizations are 7.7×10^{-4} V for 800 °C, 7.2×10^{-4} V for 750 °C, 6.8×10^{-4} V for 700 °C, and 6.5×10^{-4} V for 650 °C, respectively. Log(ΔCP) versus $\Delta i/i$ plots for different anode (**c**) and cathode (**d**) thicknesses with an operating temperature of 700 °C and an applied current density of 500 A m^{-2}. Reprinted from Ref. [26]. Copyright (2010), with permission from Elsevier

versus $\Delta i/i$ plots for anodes all increase with increasing Δi from -1 to 1. This indicates that an increased electronic conduction contribution can indeed result in increased sensitivity of polarization loss evaluation error resulted from current measurement uncertainty. The increasing trend of this sensitivity with respect to Δi appears to be more obvious at higher gas diffusivity measurement temperatures, as confirmed by the spread-out feature of the anode ΔCP versus $\Delta i/i$ plots at 650, 700, 750, and 800 °C. With the same Δi, the anode ΔCP at 700 °C appears to be larger than the anode ΔCP at 750 °C, which is somewhat similar to the crossing features observed in the previous CP evaluations [11, 42].

Figure 3.16c, d shows that the Log(ΔCP) versus $\Delta i/i$ plots for different anode (cathode) thicknesses at 700 °C. For each anode (cathode) thickness, the Log(ΔCP) versus $\Delta i/i$ plot looks similar to the Log(Δi_a) (Log(Δi_c)) versus $\Delta i/i$ plots. However, the thickness dependence of Log(ΔCP) versus $\Delta i/i$ plots appears to be opposite to the thickness dependence of the $\text{Log}(\Delta i_a)(\text{Log}(\Delta i_c))$ versus $\Delta i/i$ plots; $\text{Log}(\Delta\text{CP})$ increases with increasing the electrode thickness, whereas $\text{Log}(\Delta i_a)(\text{Log}(\Delta i_a))$ decreases with increasing the electrode thickness. The cathode ΔCP exhibits a higher thickness dependence than anode ΔCP does. For example, ΔCP reaches an infinite value as the cathode thickness exceeds 1.0 mm, while anode $\text{Log}(\Delta\text{CP})$ versus $\Delta i_a/i$ dependence does not rapidly go to infinity with an anode thickness above 7.5 mm. The opposite dependence, as shown here, suggests that for thin electrodes, the accuracy of polarization loss evaluation is less sensitive to the current measurement uncertainty compared with thick electrodes. Therefore, compared with thin electrodes, an accurate polarization loss evaluation on thick electrodes tolerates less electronic contribution of electrolytes. The opposite dependences of $\text{Log}(\Delta i_a)(\text{Log}(\Delta i_a))$ versus $\Delta i/i$ plots and $\text{Log}(\Delta\text{CP})$ versus $\Delta i/i$ plots suggest that the evaluation errors of both $\Delta i_a(\Delta i_c)$ and ΔCP cannot be reduced by reducing or increasing the electrode thickness.

3.4.2 Pressure Error

The gas diffusivity measurement error (ΔD) and the pressure uncertainty induced by gas leakage are related through Eqs. 3.22 and 3.23 [43],

$$\Delta D^{\text{eff}}_{H_2-H_2O} = \frac{RTl_ai}{4F(p^o_{H_2} - \Delta p^i_{H_2})} \tag{3.22}$$

$$\Delta D^{\text{eff}}_{O_2-N_2} = \frac{RTl_ci}{8F(\Delta p^i_{O_2} - \Delta p^o_{O_2})} \tag{3.23}$$

where $\Delta p^i_{H_2}(\Delta p^i_{H_2})$ is the $H_2(O_2)$ pressure deviation induced by the gas leak.

According to Eqs. 3.22 and 3.23, the plot of the diffusivity measurement error versus the pressure uncertainty is shown in Fig. 3.17. For both anodes and cathodes, the measurement error of gas diffusivity increases with increasing pressure uncertainty induced by gas leak. For the four considered working temperatures, 650, 700, 750, and 800 °C, the diffusivity measurement error first increases slowly with increasing pressure uncertainty and then increases abruptly as the $\Delta p/p$ approaches 0.7. This suggests that the diffusivity measurement error can become substantial since a large uncertainty in the pressure can be caused by gas leak during gas diffusivity measurement. The temperature dependence of this correlation between diffusivity measurement error and pressure uncertainty appears to be random. For instance, for 800 °C, the anode gas diffusivity measurement error is the largest among the four considered temperatures, while for 800 °C, the cathode gas diffusivity measurement error is the smallest among the four temperatures. Therefore, reducing the gas leak is necessary for the gas diffusivity measurement in the full range of the considered fuel cell operating temperatures.

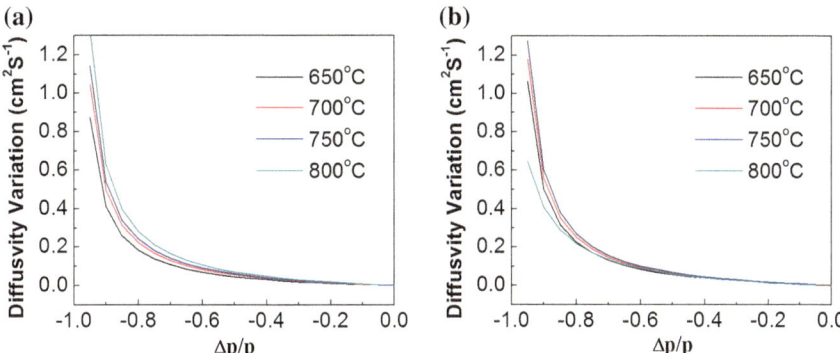

Fig. 3.17 a Plots of anode gas diffusivity measurement error versus pressure uncertainty induced by gas leak **b** plots of cathode gas diffusivity measurement error versus pressure uncertainty induced by gas leak. Anode thickness is 750 μm, and cathode thickness is 200 μm. Reprinted from Ref. [43]. Copyright (2014), with permission from International Association for Hydrogen Energy

To analyze the evaluation error of LCD, the evaluation error as a function of pressure uncertainty is plotted according to Eqs. 3.18 and 3.19 as shown in Fig. 3.18. The increase of the gas diffusivity measurement error with increasing pressure uncertainty is rather similar to the increase of the gas diffusivity measurement error with increasing pressure uncertainty, featured by a gradual increase followed by an abrupt increase as the pressure uncertainty reaches ~0.7. Figure 3.18c, d shows the plots of $\log(\Delta i_{\text{limiting}})$ versus $\Delta p/p$ for different anode and cathode thicknesses at 700 °C. With fixed $\Delta p/p$, $\log(\Delta i_a)$ increases with decreasing anode thickness; $\log(\Delta i_a)$ is the largest for 75 nm anode thickness, followed by the $\log(\Delta i_a)$ for 7.5 μm, 750 μm, and 7.5 mm anode thicknesses, respectively. Such a temperature dependence indicates that reducing the gas leak in the gas diffusivity measurement of thin anodes, such as nanostructured anodes, is of paramount importance to ensure the accurate evaluation of anode limiting current density. However, such a clear thickness dependence is absent in the plots of $\log(\Delta i_c)$ versus $\Delta p/p$ for cathodes, as suggested by the largest $\log(\Delta i_c)$ for 200 μm cathode thickness among the four considered cathode thicknesses, followed by the $\log(\Delta i_c)$ for 2.0 mm, 20 nm, and 2.0 μm cathode thicknesses.

Concentration polarization, along with Ohmic polarization and activation polarization, is an important parameter for a fuel cell system and can increase dramatically as the motion of gas species in fuel cells becomes notably impeded [44–46]. The evaluation of CP as a function of the pressure uncertainty is, thus, necessary. Based on Eqs. 3.20 and 3.21, the plot of the CP evaluation error versus pressure uncertainty is depicted in Fig. 3.19. For both anodes and cathodes, ΔCP increases with increasing pressure uncertainty for different measurement temperatures. The plot of ΔCP versus $\Delta p/p$ does not show clear temperature dependence, as indicated by the crossing feature of ΔCP versus $\Delta p/p$ plots for the four considered measurement temperatures. The cathode ΔCP increases with decreasing

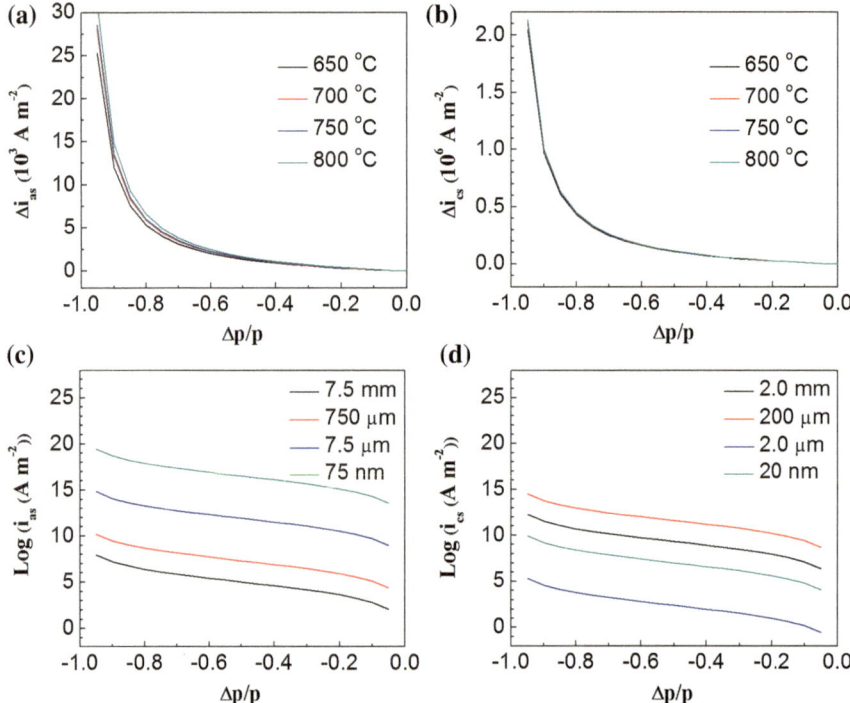

Fig. 3.18 **a** Plots of anode limiting current evaluation error versus pressure uncertainty induced by gas leak and **b** plots of cathode limiting current evaluation error versus pressure uncertainty induced by gas leak. In **a** and **b**, anode thickness is 750 μm and cathode thickness is 200 μm. **c** Plots of limiting current evaluation error in log scale versus pressure uncertainty induced by gas leak and **d** plots of cathode limiting current evaluation error in log scale versus pressure uncertainty induced by gas leak. In **c** and **d**, the measurement temperature is 700 °C. Reprinted from Ref. [43]. Copyright (2014), with permission from International Association for Hydrogen Energy

measurement temperature with fixed $\Delta p/p$. This temperature dependence suggests that eliminating the gas leak is particularly important for low-temperature fuel cells.

Comparing ΔCP versus $\Delta p/p$ plots for anodes and cathodes, the ΔCP for cathodes is much more sensitive to the pressure uncertainty induced by gas leak than is the ΔCP for anodes. This suggests that the gas leak created during the gas diffusivity measurement of porous cathodes can, indeed, result in a particularly high inaccuracy in the CP evaluation of fuel cells. Compared to the gradual increasing slopes of ΔD versus $\Delta p/p$ and $\Delta i_{\text{limiting}}$ versus $\Delta p/p$ plots as $\Delta p/p < 0.7$, ΔCP increases more abruptly with increasing $\Delta p/p$. The evaluation of the concentration polarization is thus more sensitive to the pressure uncertainty induced by the gas leak in the electrochemical devices for gas diffusivity measurement.

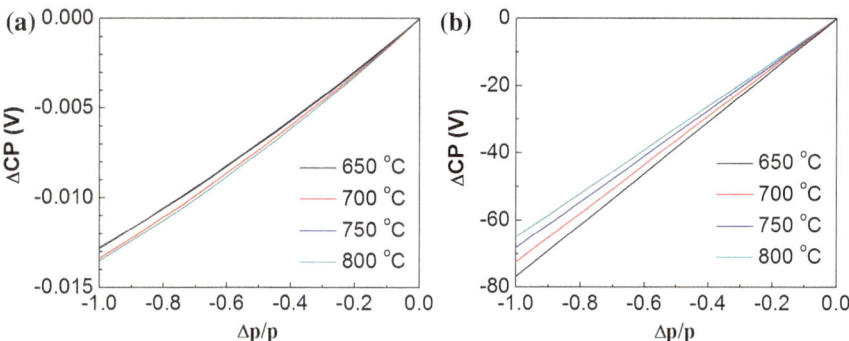

Fig. 3.19 a Plots of anode concentration polarization evaluation error versus pressure uncertainty induced by gas leak and **b** plots of cathode concentration polarization evaluation error versus pressure uncertainty induced by gas leak. In **a** and **b**, anode thickness is 750 μm and cathode thickness is 200 μm. Reprinted from Ref. [43]. Copyright (2014), with permission from International Association for Hydrogen Energy

3.4.3 Temperature Error

The evaluation error ΔD induced by the temperature variance ΔT can be calculated with known applied current density and electrode thickness, based on Eqs. 3.24 and 3.25 [47].

$$\Delta D_{H_2-H_2O}^{\text{eff}} = \frac{Rl_a i}{4F(p_{H_2}^o - p_{H_2}^i)}\Delta T \tag{3.24}$$

$$\Delta D_{O_2-N_2}^{\text{eff}} = \frac{Rl_c i}{8F(p_{O_2-N_2}^i - p_{O_2-N_2}^o)}\Delta T \tag{3.25}$$

As shown in Fig. 3.20, for both anodes (Fig. 3.20a) and cathodes (Fig. 3.20b), ΔD increases linearly with increasing ΔT as $\Delta T/T$ varies in the range of 0–20 %. The relatively steep slopes of the ΔD plots for different temperatures suggest that an uncertainty in the temperature indeed induces substantial errors in the gas diffusivity evaluation. With a certain temperature uncertainty, for both electrodes, the temperature uncertainty for high-temperature measurements results in larger evaluation errors compared with the evaluation errors for low-temperature measurement. With a fixed value of temperature uncertainty, the error of the diffusivity evaluation for measurement temperature 800 °C is the largest among the four considered measurement temperatures, followed by the errors of the diffusivity evaluation for measurement temperatures 750, 700, and 650 °C, respectively, as shown in Fig. 3.20a, b. Thus, the temperature uncertainty instills more concern for the gas diffusivity evaluation of high-temperature fuel cells. The proposed device is highly

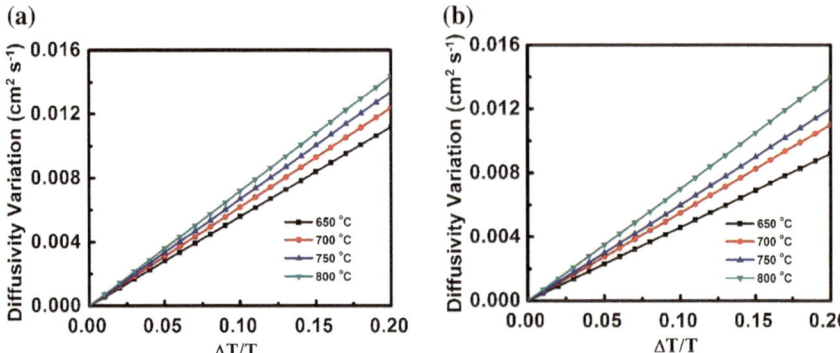

Fig. 3.20 Plots of anode (**a**) and cathode (**b**) diffusivity measurement errors as a function of the uncertainty in measurement temperature ($\Delta T/T$) with an applied current density of 500 A m^{-2}. Anode hydrogen effective binary diffusivities are 0.046 cm^2 s^{-1} for 650 °C, 0.055 cm^2 s^{-1} for 700 °C, 0.06 cm^2 s^{-1} for 750 °C, and 0.07 cm^2 s^{-1} for 800 °C, respectively. Cathode oxygen effective binary diffusivities are 0.056 cm^2 s^{-1} for 650 °C, 0.062 cm^2 s^{-1} for 700 °C, 0.067 cm^2 s^{-1} for 750 °C, and 0.072 cm^2 s^{-1} for 800 °C, respectively. Anode thickness is 0.75 mm, and cathode thickness is 0.2 mm. Reprinted from Ref. [47]. Copyright (2014), with permission from Elsevier

desirable for SOFCs since the working temperatures of currently existing SOFCs are typically above 500 °C.

As shown in Eqs. 3.18 and 3.19, the limiting current is quantitatively correlated with the gas diffusivity. Hence, we evaluate the error in the evaluation of limiting current density as a function of temperature uncertainty in the gas diffusivity measurement of fuel cells. $\Delta i_a (\Delta i_c)$ is linearly increasing with increasing $\Delta T/T$ as shown in Fig. 3.21. For anodes, Δi_a appears to be less dependent on temperature in the medium temperature range compared to the Δi_a at low and high temperatures, as noted by the close magnitudes of Δi_a at 700 and 750 °C. Δi_c increases sharply with increasing $\Delta T/T$ at any temperature as noted by the almost overlapping plots of Δi_c versus $\Delta T/T$, as shown in Fig. 3.21b. Unlike the monotonous increase in the evaluation error of limiting current density with increasing measurement temperature, the evaluation error of cathode limiting current density exhibits an irregular dependence on electrode thickness although $\Delta i_a (\Delta i_c)$ increases with increasing $\Delta T/T$ for all the considered electrode thicknesses in Fig. 3.21c, d. For instance, with a fixed temperature uncertainty, the evaluation error of limiting current density is the largest for 0.2 mm cathodes, followed by 2 mm, 20 nm and 2 μm cathodes, respectively. However, no irregularity is observed in the Log(Δi_a) versus $\Delta T/T$ plots for anodes with various thicknesses at 700 °C. The irregular cathode thickness dependence suggests that eliminating the temperature uncertainty in the diffusivity measurement in fuel cells is necessary for any cathode thickness and that the effort of achieving accurate gas diffusivity measurement via the previous electrochemical devices by selecting certain electrode thicknesses is not viable.

Both diffusivity and limiting current interplay quantitatively with the concentration polarization of fuel cells, a major source of polarization in fuel cells

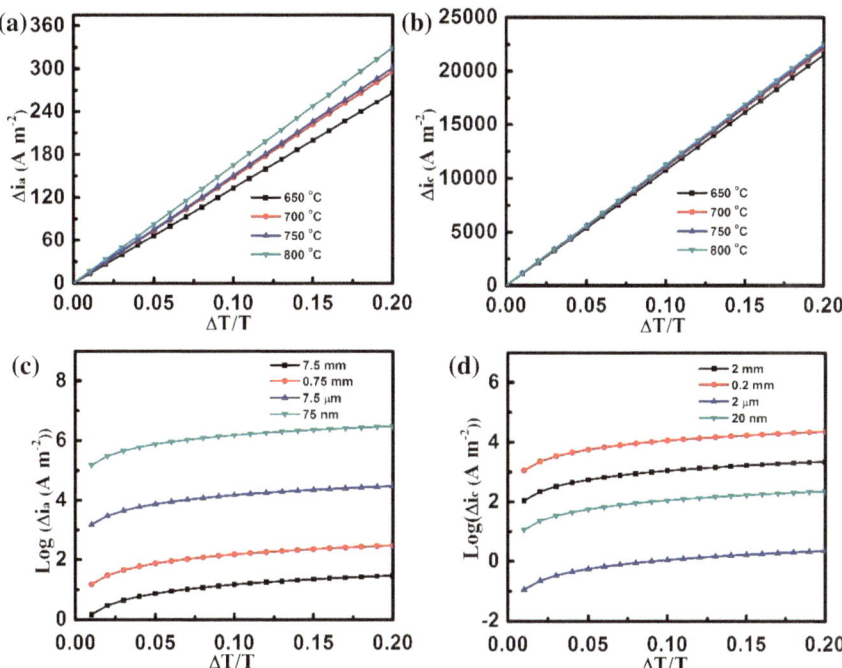

Fig. 3.21 Plots of the evaluation errors in anode (**a**) and cathode (**b**) limiting current density as a function of the measurement temperature uncertainty ($\Delta T/T$) with an applied current density of 500 A m^{-2}. Anode thickness is 0.75 mm, and cathode thickness is 0.2 mm. Anode limiting current densities are 1.33×10^3 A m^{-2} for 650 °C, 1.48×10^3 A m^{-2} for 700 °C, 1.51×10^3 A m^{-2} for 750 °C, and 1.65×10^3 A m^{-2} for 800 °C, respectively. Cathode limiting current densities are 10.77×10^4 A m^{-2} for 650 °C, 11.07×10^4 A m^{-2} for 700 °C, 11.17×10^4 A m^{-2} for 750 °C, and 11.25×10^4 A m^{-2} for 800 °C, respectively. Log(Δi_a) (Log(Δi_c)) versus $\Delta T/T$ plots for different electrode thicknesses at 700 °C and with an applied current density of 500 A m^{-2}. Reprinted from Ref. [47]. Copyright (2014), with permission from Elsevier

with electrodes characterized by impeded gas transport [27–29]. To evaluate the necessity of the multi-functional sensor electrochemical device as proposed in this report, the concentration polarization as a function of temperature uncertainty was analyzed at different measurement temperatures and for electrodes of different thicknesses. ΔCP increases with increasing temperature uncertainty in all the ΔCP versus $\Delta T/T$ plots for anodes and cathodes at different temperatures and for different electrode thicknesses, as shown in Fig. 3.22. Interestingly, unlike the plots of diffusivity and limiting current density evaluation errors in Figs. 3.20 and 3.21, the concentration polarization evaluation error exhibits an irregular dependence on the measurement temperature for anodes. For instance, with a fixed temperature uncertainty, the evaluation error of the anode concentration polarization at 650 °C is the largest among the four considered temperatures, followed by the evaluation error of anode concentration polarization at 750, 800, and 700 °C, respectively, as shown in Fig. 3.22a. Further, irregular electrode thickness

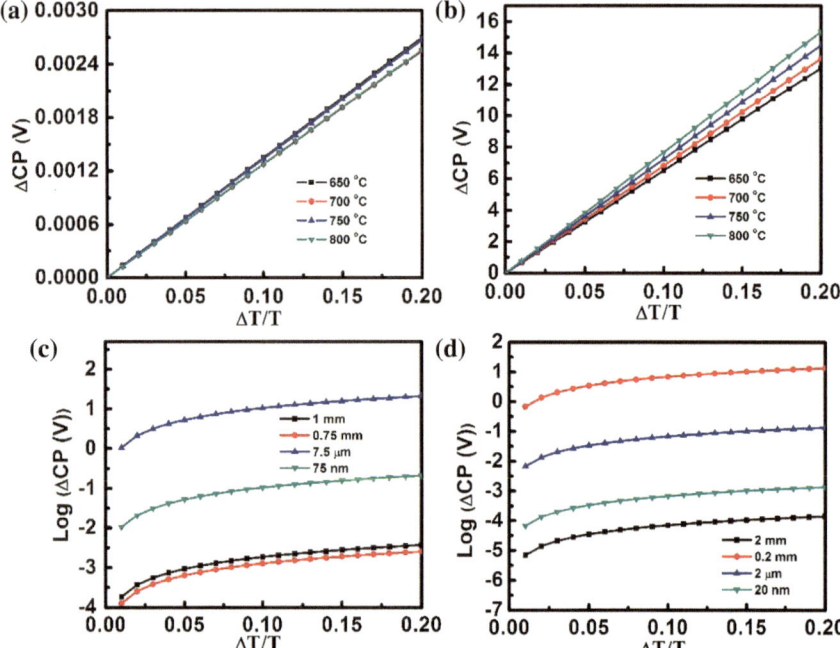

Fig. 3.22 Plots of the evaluation errors in the anode (**a**) and cathode (**b**) concentration polarization as a function of the uncertainty ($\Delta T/T$) in measurement temperature. Anode concentration polarizations are 0.0128 V for 650 °C, 0.0134 V for 700 °C, 0.0128 V for 750 °C, and 0.0135 V for 800 °C, respectively. Cathode concentration polarizations are 6.5×10^{-4} V for 650 °C, 6.8×10^{-4} V for 700 °C, 7.2×10^{-4} V for 750 °C, and 7.7×10^{-4} V for 800 °C, respectively. Log(ΔCP) versus $\Delta T/T$ plots of anode (**c**) and cathode (**d**) for different electrode thicknesses at 700 °C and with an applied current density of 500 A m^{-2}. Anode thickness is 0.75 mm, and cathode thickness is 0.2 mm. Reprinted from Ref. [47]. Copyright (2014), with permission from Elsevier

dependence is also observed in the Log(ΔCP) versus $\Delta T/T$ plots for different anode (cathode) thicknesses at 700 °C. The irregular dependences of concentration polarization on the measurement temperature and the electrode thicknesses further confirm the necessity of eliminating the temperature uncertainty in the gas diffusivity measurement in fuel cells. Therefore, the proposed electrochemical device is highly desirable for gas diffusivity measurements in both bulk and thin nanostructured fuel cell systems at any feasible operating temperature.

References

1. Z. Yu, R.N. Carter, J. Power Sources **195**, 1079–1084 (2010)
2. C. Chan, N. Zamel, X. Li, J. Shen, Electrochim. Acta **65**, 13–21 (2012)
3. J. Crank, *The Mathematics of Diffusion*, 2nd cdn. (Oxford University Press, NY, 1975). (Edmonton, Alberta, Canada, 2011)

4. T. Marrero, E. Mason, J. Phys. Chem. Ref. Data **1**(1), 3–118 (1972)
5. L.M. Pant, S.K. Mitra, M. Secanell, J. Power Sources **206**, 153–160 (2012)
6. L.M. Pant, S.K. Mitra, M. Secanell, in: *ASME 2011 9th International Conference on Nanochannels, Microchannels, and Minichannels*, no. ICNMM2011-58181, Edmonton, Alberta, Canada, 2011
7. W. He, K.J. Yoon, R.S. Eriksen, S. Gopalan, S.N. Basu, U.B. Pal, J. Power Sources **195**, 532 (2010)
8. F. Zhao, T.J. Armstrong, A.V. Virkar, J. Electrochem. Soc. **150**, A249–A256 (2003)
9. W. He, B. Wang, H. Zhao, Y. Jiao, J. Power Sources **196**, 9985 (2011)
10. W. He, B. Wang, Adv. Energy Mater. **2**, 329–333 (2012)
11. W. He, B. Wang, J.H. Dickerson, Overall concentration polarization and limiting current density of fuel cells with nanostructured electrodes. Nano Energy **1**, 828–832 (2012)
12. E.A. Wargo, A.C. Hanna, A. Cecen, S.R. Kalidindi, E.C. Kumbur, J. Power Sources **197**, 168–179 (2012)
13. J.R. Wilson et al., Nat. Mater. **5**, 541–544 (2006)
14. A. Cecen et al., J. Electrochem. Soc. **159**, B299–B307 (2012)
15. L. Cindrella et al., J. Power Sources **194**, 146–160 (2009)
16. A. Rowe, X.G. Li, J. Power Sources **102**, 82–96 (2001)
17. S.H. Chan, C.F. Low, O.L. Ding, J. Power Sources **103**, 188–200 (2002)
18. J.B. Ge, A. Higier, H.T. Liu, J. Power Sources **159**, 922–927 (2006)
19. D.M. Bernardi, M.W. Verbrugge, AIChE J. **37**, 1151–1163 (1991)
20. M.L. Perry, J. Newman, E.J. Cairns, J. Electro-Chem. Soc. **145**, 5–15 (1998)
21. A.S. Joshi, K.N. Grew, A.A. Peraccio, W.K.S. Chiu, J. Power Sources **164**, 631–638 (2007)
22. W. He, X. Lin, J.H. Dickerson, J.B. Goodenough, An electrochemical device for three-dimensional (3D) diffusivity measurement in fuel cells. Nano Energy **2**(5), 1004–1009 (2013)
23. D. Ding, L. Li, K. Feng et al., J. Power Sources **187**, 400–402 (2009)
24. J.W. Kim, A.V. Virkar, K.Z. Fung, K. Metha, S.C. Singhal, J. Electrochem. Soc. **146**, 69 (1999)
25. E.L. Cussler, *Diffusion: Mass Transfer in Fluid Systems* (Cambridge University Press, Cambridge, UK, 1995)
26. W. He, B. Wang, A current-sensor electrochemical device for accurate gas diffusivity measurement in fuel cells. J. Power Sources **232**, 93–98 (2013)
27. Z. Lin, G. Waller, Y. Liu, M. Liu, C.P. Wong, Adv Energy Mater. **2**, 884–888 (2012)
28. T. Howells, E. New, P. Sullivan, T.S. Jones, Adv. Energy Mater. **1**, 1085–1088 (2011)
29. J.R. Moore, S. Albert-Seifried, A. Rao, S. Massip, B. Watts, D.J. Morgan, R.H. Friend, C.R. McNeill, H. Sirringhaus, Adv. Energy Mater. **1**, 23–240 (2011)
30. H.J. Lee, H. Strathmann, S.H. Moon, Desalination **190**, 43–50 (2006)
31. J. Santamaria, J. Garcia-Barriocanal, A. Rivera-Calzada, M. Varela, Z. Sefrioui, E. Iborra, C. Leon, S.J. Pennycook, Science **321**, 676–680 (2008)
32. H. Wu, G. Chan, J.W. Choi, I. Ryu, Y. Yao, M.T. McDowell, S.W. Lee, A. Jackson, Y. Yang, L. Hu, Y. Cui, Nat. Nanotechnol. **7**, 310–315 (2012)
33. H.J. Choi, S.M. Jung, J.M. Seo, D.W. Chang, L. Dai, J.B. Baek, Nano Energy **1**, 534–551 (2012)
34. M. Liu, Y.M. Choi, L. Yang, K. Blinn, W. Qin, P. Liu, M.L. Liu, Nano Energy **1**, 448–455 (2012)
35. M. Zhang, L. Dai, Nano Energy **4**, 514–517 (2012)
36. J. Suffner, S. Kaserer, H. Hahn, C. Roth, F. Ettingshausen, Adv. Energy Mater. **1**, 648–654 (2011)
37. B. Tu, W. Li, F. Zhang, Y.Q. Dou, Z.X. Wu, H.J. Liu, X.F. Qian, D. Gu, Y.Y. Xia, D.Y. Zhao, Adv. Energy Mater. **1**, 382–386 (2011)
38. J.M. Serra, I. Garcia-Torregrosa, M.P. Lobera, C. Solis, P. Atienzar, Adv. Energy Mater. **1**, 618–625 (2011)
39. D. Zhao, J.H. Li, M.K. Song, B.L. Yi, H.M. Zhang, M.L. Liu, Adv. Energy Mater. **1**, 203–211 (2011)
40. L. Fan, C. Wang, B. Zhu, Nano Energy **1**, 631–639 (2012)
41. L. Su, Y.X. Gan, Nano Energy **1**, 159–163 (2012)

42. W.D. He, B. Wang, J.H. Dickerson, Nano Energy **1**, 828–832 (2012)
43. W. He, X. Lin, W. Lv, J.H. Dickerson, Electrochemical devices with optimized gas tightness for the diffusivity measurement in fuel cells. Int. J. Hydrogen Energy **39**(5), 2334–2339 (2014)
44. F. Zhao, A.V. Virkar, Dependence of polarization in anode-supported solid oxide fuel cells on various cell parameters. J. Power Sources **141**, 79–95 (2005)
45. T.J. Yen, N. Fang, X. Zhang, G.Q. Lu, C.Y. Wang, A micro methanol fuel cell operating at near room temperature. Appl. Phys. Lett. **83**, 4056–4058 (2003)
46. L.A. Chick, K.D. Meinhardt, S.P. Simner, B.W. Kirby, M.R. Powell, N.L. Canfield, Factors affecting limiting current in solid oxide fuel cells or debunking the myth of anode diffusion polarization. J. Power Sources **196**, 4475–4482 (2011)
47. W. He, J.B. Goodenough, An electrochemical device with a multifunctional sensor for gas diffusivity measurement in fuel cells. J. Power Sources **251**, 108–112 (2014)

Chapter 4
Solid Oxide Fuel Cells with Improved Gas Transport

4.1 Introduction

Solid oxide fuel can utilize various fuels, including hydrogen, natural gas, and even traditional fossil fuel, to generate electricity with a conversion efficiency larger than 60 % [1–4]. Such advantages make SOFCs an environmental-friendly approach toward meeting the clean-energy demands to our society. The main issue of SOFCs is the high operation temperature, resulting in high cost and material challenges, such as poor gas transport through the cells [5, 6]. By developing high ionic conductive electrolyte materials at intermediate and low temperatures, the ohmic polarization can be greatly decreased. Both anode and cathode materials with high electro-catalytic properties are also in high demand to decrease the activation polarization. Nevertheless, a huge obstacle to further improve the performance of SOFCs at low temperatures is their high concentration polarization associated with both anodes and cathodes, especially under high-current-density operations. The main methodology to improve fuel gas transport in anodes, oxygen gas transport in cathodes, and the values for the concentration polarization is to achieve high-performance electrode microstructures by improving the porosity, pore size, tortuosity and surface area. In this chapter, material advancements in SOFC electrodes are first reviewed since enhanced gas transport is based on the specific characteristics of the electrode materials, such as superior catalytic activity. The synthesis methodologies for the microstructure control of these highly catalytic-active electrode materials, including porosity, pore size and tortuosity, are discussed, followed by the overview of the newly developed characterization techniques for analyzing the electrode microstructures. Finally, the correlation of the microstructure of SOFC electrodes with the gas diffusion performance is illustrated in detail.

© The Author(s) 2014   45
W. He et al., *Gas Transport in Solid Oxide Fuel Cells*, SpringerBriefs in Energy,
DOI 10.1007/978-3-319-09737-4_4

4.2 Brief Review of SOFC Electrode Materials

SOFCs typically follow a sandwich structure comprising a solid oxide electrolyte flanked by an anode on one side and a cathode on the other side. The solid oxide electrolyte plays a dual role, separating the fuel gas (such as H_2) and air, facilitating the conduction of O^{2-} from the cathode to the anode. The electrolyte is the key component of an SOFC, which, to a large extent, determines the selection of both the anode and cathode materials. The most commonly used electrolyte material is yttria-stabilized zirconia (YSZ), a material that has been employed since the initial stages of SOFC development. YSZ is relatively inexpensive and is chemically and thermo-mechanically stable. At 973 K, the ohmic loss associated with a 10 μm YSZ electrolyte is only 0.05 Ω cm^2, and the performance remains stable after extended periods of operation [7]. Anodes are typically porous composites of Ni and YSZ. Ni provides electronic conductivity and catalytic activity, while YSZ provides channels for ionic conduction within the electrode, helps maintain porosity, and helps match the coefficients of thermal expansion between the electrode and electrolyte. Cathodes must maintain profound electronic conductivity at high temperatures in air, and the most common component is a conductive oxide, Sr-doped $LaMnO_3$ (LSM) [8]. Similar to the anode, LSM is often mixed with YSZ to form a cathode composite.

The aforementioned traditional SOFC setup, Ni-YSZ (anode)|YSZ (electrolyte)|LSM (cathode), suffers from high operating temperatures due to the high activation barrier for ionic conduction in YSZ, subsequently leading to higher system costs and greater performance degradation rates. Anode-supported cells with significantly thinner YSZ were developed to minimize the ohmic polarization of electrolyte and to reduce the operating temperature. For instance, to retain the same ohmic resistance of 0.15 Ω cm^{-2}, ~150-μm YSZ can meet the requirement at 950 °C, but the thickness of YSZ must be reduced to ~1 μm at 500 °C [9]. Conventional ceramic processing (e.g., tape casting), however, can only realize films with a minimum thickness of ~10 μm, which limits the operating temperature to a range of values >700 °C for YSZ. Therefore, electrolytes of higher ionic conductivity at lower temperatures (<700 °C) are needed. Various alternative electrolytes have been developed, among which aliovalent-doped ceria, isovalent-cation-stabilized bismuth, and perovskite-structured doped $LaGaO_3$ are particularly attractive due to their superior ionic conductivity. For instance, at 500 °C, the area-specific resistance of 10-μm thick YSZ, SDC ($Ce_{0.9}Gd_{0.1}O_{1.95}$), SDC ($Ce_{0.9}Sm_{0.1}O_{1.95}$), ESB ($Er_{0.4}Bi_{1.6}O_3$), and LSGM ($La_{0.9}Sr_{0.1}Ga_{0.8}Mg_{0.2}O_3$) are 1.259, 0.143, 0.402, 0.037, and 0.181 Ω cm^{-2}, respectively [10, 11]. For different electrolytes, various catalytically active anodes and cathodes are developed to match the specific electrolyte with good chemical and thermal compatibility. These anode and cathode materials will be discussed in detail in the following paragraphs.

The general requirements for SOFC anode materials should include high electronic conductivity as well as ionic conductivity, high catalytic activity toward the oxidation of fuels, high chemical and thermal stability, high compatibility with the electrolyte, facile fabrication, and low cost [12]. Metal-based cermet, such as

Ni-YSZ, Ni-GDC, Ni-SDC, and Ni-LSGM, is normally used as anode material since they satisfy most of the above requirements. The same electrolyte components in the anode cermet ensure chemical and thermal compatibility and superior anode–electrolyte interfaces. Another role of the electrolyte components is to expand the triple-phase boundaries (TPBs) to bulk-anode regions far from the electrolyte–anode interfaces. Ni is dispersed homogenously in the anode cermet to provide high catalytic activity as well as porous structures for low fuel gas diffusion resistance. The main drawbacks of Ni-based cermet is their poor tolerance to carbon deposition and sulfur poisoning when hydrocarbon fuels that contain trace impurity of H_2S are used. Replacing Ni with Cu can prevent carbon deposition but will significantly decrease the catalytic activity [13]. The addition of small amounts of highly catalytically active noble metal components, such as Pd and Ru, has confirmed the potential in the utilization of fossil gas fuels in SOFCs. More detailed information in anode materials can be found in the literature [14–18].

A cathode in SOFCs electro-reduces O_2 to O^{2-} and drives O^{2-} toward the electrolyte. The general requirement for cathode materials is similar to that for anode materials and includes superior mixed ionic and electron conductivity (MIEC), high electro-reduction activity, good chemical and thermal stability and compatibility with the electrolyte. The main candidates for use as cathode materials are perovskite oxides, such as LSM ($La_{0.8}Sr_{0.2}MnO_{3-\delta}$), LSCF ($La_{0.6}Sr_{0.4}Co_{0.2}Fe_{0.8}O_{3-\delta}$), SSC ($Sm_{0.5}Sr_{0.5}CoO_{3-\delta}$), and BSCF ($Ba_{0.5}Sr_{0.5}Co_{0.8}Fe_{0.2}O_{3-\delta}$). As mentioned before, LSM is the most widely used cathode and remains the practical choice for operation in 700–1,000 °C due to its superior long-term chemical stability and compatibility with YSZ, GDC, SDC, and LSGM. Nevertheless, low ionic conductivity (only 10^{-6} times that of YSZ), electronic conductivity at low temperatures, and low catalytic activity limits its use in low-temperature SOFCs (LT-SOFCs). Cobalt-containing perovskites have high MIEC and exceptional electro-reduction activity. Unfortunately, they typically have high thermal expansion coefficients (TECs) and react with YSZ at temperatures as low as 700 °C to form insulating phases. Partial substitution of Co with Fe decreases TECs dramatically, resulting in the thermal compatibility with doped ceria electrolyte (GDC, SDC) and LGSM [19]. LSCF and BSCF are other such cathodes that perform well in the operation temperature of 500–700 °C [20–22]. Another cobalt-containing perovskite, SSC, also shows exceptional performance in ceria or LSGM-based SOFCs [23]. These MIEC cathodes expand the TPBs in the bulk cathode rather than just at the cathode-electrolyte interfaces and improve SOFCs performances at low operating temperatures. More detailed information of cathode materials can be found from previous excellent reviews [24–27].

Despite the catalytic and compatibility issues, both the fuel gas diffusion in anodes and air diffusion in the cathode structures are of great significance since they normally causes severe concentration polarization energy loss especially under large current operation. Therefore, a catalytically active anode or cathode material with desired microstructure is fundamentally important to improve the gas diffusions. Based on the electrode material development of SOFCs, the microstructure control, characterization, and SOFC performance will be the focus in the following sections.

4.3 Synthesis Methodology for Microstructure Control of SOFC Electrodes

To assemble a single cell comprising a thin film electrolyte, the electrolytic cell can be cast onto the supporting anode by a number of techniques, like extrusion, tape casting, screen printing, spin coating, or another effective method [28–33]. These green cells are then sintered at high temperatures to allow for densification of the electrolyte layer. The cathode is deposited onto the sintered electrolyte surface through brushing, dip-coating, or other techniques. This is followed by firing the material at an elevated temperature to make an adequate connection between the electrode particles and to obtain good adhesion between the electrode layer and the electrolyte surface. This minimizes the contact resistance and ensures the sufficient mechanical strength of the electrode layer. In the above fabricating process, sufficient porosity, and rational microstructures for both anode and cathode are needed for the free transportation of oxygen, fuels and reaction products within the electrode channels. Otherwise, a serious concentration polarization could occur, especially at high polarization currents. Many methods are developed to control the porosity and tune the microstructures of SOFC electrodes.

One of the main synthetic methods for porous structured SOFC electrodes is the traditional ceramic or cermet sintering process. A porous Ni-YSZ anode can be synthesized by high-temperature sintering of the mixture of NiO, YSZ, and a pore former (such as graphite, carbon spheres, polymer, and starch), followed by a high-temperature hydrogen reduction of NiO to Ni. The resulting pore size is determined by the size of NiO particle in the cermet. The porosity of the Ni-YSZ anode was controlled by the amount of NiO and pore former added [34, 35]. The porosity of cathodes can also be varied by varying the amount of pore formers in the slurries. For instance, the porosity of LSM film can be tuned between 29 and 48 vol% as the carbon spheres added vary from 20 to 30 wt% carbon in the ceramic slurry. The pore size is determined by the size of carbon spheres after the ball-milling of the slurry.

Syntheses have been designed to fabricate graded electrode structures, such as varying the porosity gradient by decreasing the porosity and varying the pore size from the electrode surface to the electrode–electrolyte interface. These graded-structured electrodes may allow less hindered gas transportation and may help to establish rich electrochemical active three-phase boundaries (TPBs). Several processes have been pursued to fabricate gradient anode substrates, including multilayer printing (electrolyte supported), [36] multi-step die pressing [37], freeze tape casting [38, 39], and multilayer tape casting [40]. Chen et al. developed a graphite-assisted phase-inversion process to fabricate tubular anodes with continuous porosity gradient [41]. The graphite layer plays a role in controlling the phase-separation reaction in the ceramic layer and removing the skin layer. The SEM images of the cross sections of this Ni-YSZ anode and the SEM images of anode outer surface structure far from the electrolyte with and without skin graphite surface are shown in Fig. 4.1.

Pore formers, such as graphite, polymer or carbon spheres, and starch, usually are employed in particulate form and, therefore, generate template-separated pores, as shown in Fig. 4.2a. The low pore connectivity restrains gas diffusion through

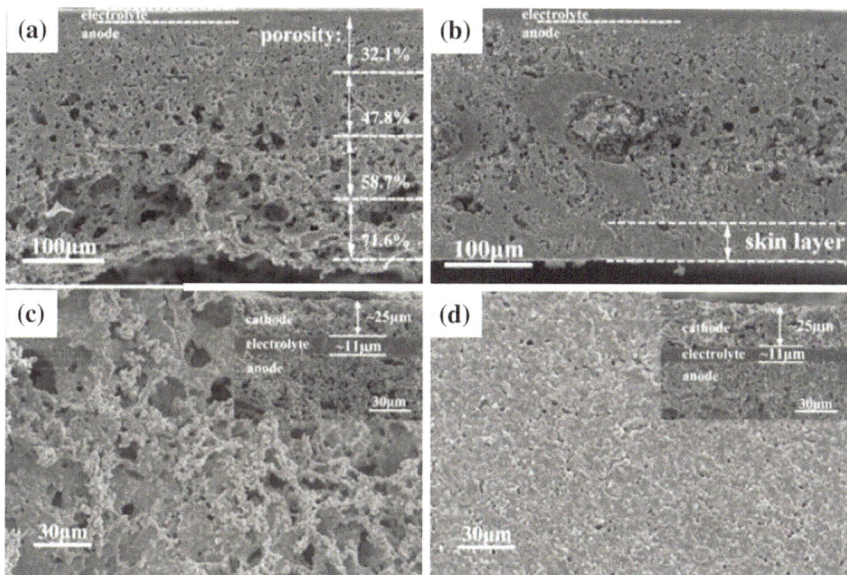

Fig. 4.1 SEM micrographs for cross-sectional microstructure of the anode **a** with pore gradient, **b** with the skin layer, and the anode outer surface far from the electrolyte **c** without skin layer, **d** with the skin layer. *Insets* of **c** and **d** depict the cross-sectional view of the electrode–electrolyte interface of the corresponding cells. Reprinted from Ref. [41]. Copyright (2014), with permission from Elsevier

Fig. 4.2 Schematic representations of the formation of porous structures by **a** using conventional particulate pore former and **b** a mesh-assisted phase-inversion process. Reprinted from Ref. [43]. Copyright (2014), with permission from Elsevier

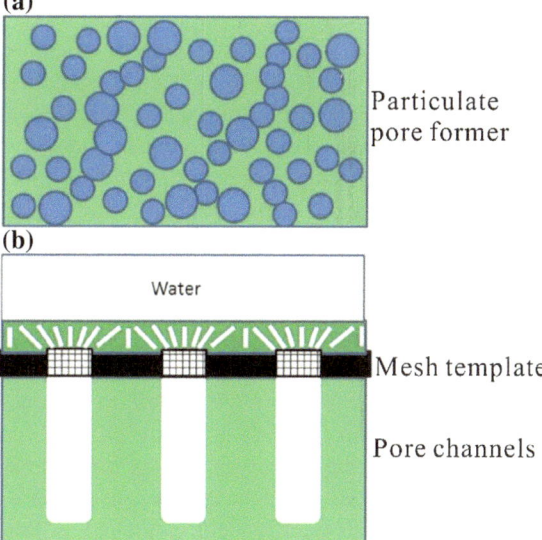

these pores. Adding more pore formers can increase the pore connectivity, but can decrease the mechanic strength of the ceramic electrode support [42]. A mesh-assisted phase-inversion process was developed to prepare micro-channeled ceramic

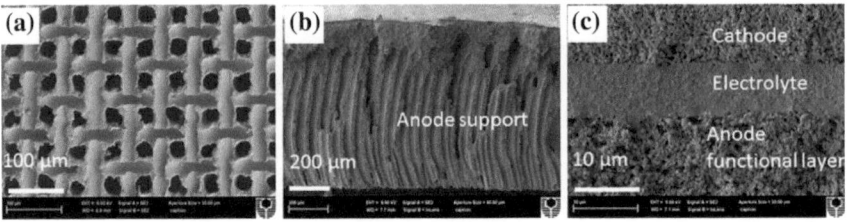

Fig. 4.3 SEM images of **a** anode surface and **b** cross section and **c** the sandwich structure of SOFC. Reprinted from Ref. [43]. Copyright (2014), with permission from Elsevier

films with enhanced pore connectivity [43, 44]. As shown in Fig. 4.2b, a slurry containing NiO/GDC and other organic additives can be cast to form a disk-shaped anode green body (in green color), and a stainless steel mesh can reside just below the slurry's surface (in black color). Water applied on the top of slurry induces phase inversion in the system because of convection of the solvent and the coagulant [45, 46]. A hierarchically micro-channel structure forms during this phase inversion: finger-like pores embed within the membranes, and sponge-like pores distribute around the finger-like pores [47, 48]. The finger-like pores act as gas diffusion channels where gas is delivered and distributed into the layer with the sponge-like pores, which acts as a functional layer, for example for catalytic reactions.

After the phase conversion, the steel mesh is extracted to remove the skin layer. The SEM image of the anode surface in Fig. 4.3a clearly shows that in the presence of uniform open channels have diameters similar to the template mesh's aperture size. The SEM image of the cross section of the anode shown in Fig. 4.3b displays that the channels cross the anode thickness and end before reaching the other side of the anode, forming a conventional porous layer acting as a functional layer to support a thin-film electrolyte and electro-catalytic area. The cross-sectional SEM image of the less porous anode functional layer, with a thin dense electrolyte and a porous cathode is shown in Fig. 4.3c. The cell test results for this micro-channeled anode-supported SOFC show that gas transport is improved significantly, and no obvious concentration polarization is observed.

As discussed above, in the phase-conversion process, a homogeneous slurry of electrode is immersed in a coagulant (non-solvent), which is miscible with the solvent in the slurry but immiscible with the polymer. Phase instability of the solution is induced by counter diffusion between the solvent and non-solvent, leading to phase separation to form an electrode film with asymmetric structure. Therefore, the phase-conversion method is clearly a facile, scalable, and cost-effective technique in tailoring the microstructures of both anode and cathode, to form graded-porosities and micro-channels with improved gas transport. The phase-conversion method already has been widely employed to fabricate SOFCs not only within a planar configuration but also within a tubular configuration [49–60]. More are expected for this method to improve the gas transport performance as well as the electro-catalytic activity of both anodes and cathodes of SOFCs if the obstacle of gas diffusion of the commonly existing dense skin layer is addressed.

4.4 Characterization Techniques of Microstructures of SOFC Electrodes

The diffusion performance of electrode microstructures of SOFCs depends on the porosity, tortuosity, pore size, and surface area of these systems, among other characteristics. The accurate measurement of these parameters largely facilitates the quantitative understanding of gas diffusion processes in electrodes and concentration polarizations of SOFCs. The Brunauer–Emmett–Teller (BET) method and mercury porosimetry are powerful tools to characterize the surface area, porosity, and pore size of SOFC electrodes [61–63]. The pore size and porosity also can be directly acquired from SEM and TEM images [64, 65]. Nevertheless, most common techniques for microstructure analysis only obtain 2D images of the electrodes. This makes obtaining accurate structure parameter information very challenging due to the lack of information regarding how electrode materials interconnect in 3D space. For instance, the evaluation of the tortuosity is almost impossible using only 2D SEM and TEM images. Up to now, several imaging techniques have been developed to investigate the three-dimensional structure of materials. Focused ion beam–scanning electron microscopy (FIB-SEM) [66–76] and full-field transmission X-ray microscopy (TXM) [71, 77–82] are two representative techniques with wide applications that provide significant insight into diffusion performance through the properties of the electrode microstructure. A brief introduction to these two techniques and their application in the characterization of SOFC electrode structures are given below.

The combined capabilities of a dual beam FIB-SEM can be employed to form 3D reconstructions of the structure of SOFCs. A schematic diagram of the FIB-SEM setup is shown in Fig. 4.4a. To image the multilayered structure, first a trench must be milled by the FIB, as shown in Fig. 4.4b, to reveal the underlying components. Thin

Fig. 4.4 **a** Schematic diagram showing the FIB–SEM geometry and **b** a low-magnification SEM image of a FIB-etched region at the anode/electrolyte interface of a SOFC. **c** A view of the 3D reconstruction showing the Ni (*green*), YSZ (*translucent/gray*), and pore (*blue*) phases. **d** A finite-element mesh converted from the pore subdomain in (**c**). Reprinted from Ref. [66]. Copyright (2006), with permission from Nature Publication Group

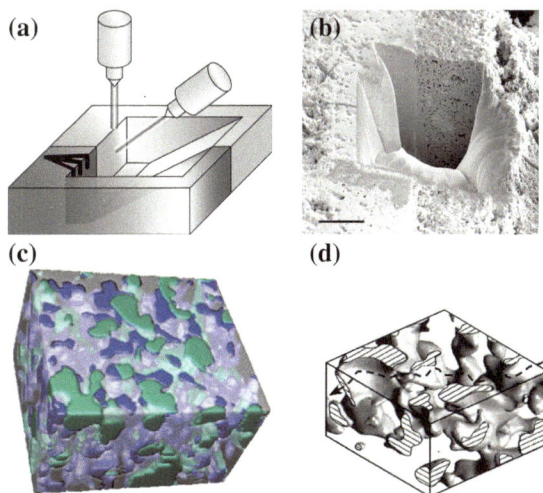

(50 nm) sections are then removed from an exposed surface by the FIB, followed by SEM imaging of the surface. This process is repeated to yield a series of consecutive SEM images. By stacking these 2D SEM images in 3D space, a 3D reconstruction of the Ni-YSZ anode can be assembled as shown in Fig. 4.4c [66]. In the 3D image, different materials are indicated by different colors with Ni in green, YSZ in translucent/gray and pore in blue. By carefully dealing with the 3D graphic data, quantitative data for several critical parameters of the anode microstructure can be obtained, including Ni and YSZ volumetric fraction, porosity, surface area, pore size, and triple-phase boundary length. The tortuosity is also calculated mathematically by converting the 3D reconstruction of the gas pores in Fig. 4.4c into a finite-element mesh as shown in Fig. 4.4d. Although FIB-SEM allows for high-resolution imaging (from 10 to 100 nm) and is widely applied in the 3D imaging of SOFC materials, the imaging time increases significantly as the sampling volume increases. A time-economic imaging volume size (around $10 \times 10 \times 10 \ \mu m^3$) with acceptable resolution (<35 nm) is far smaller than the minimum representative volume element (RVE) (around $35 \times 35 \times 35 \ \mu m^3$) in order to provide reliable statistics on effective transport properties [72, 83, 84].

TXM technique was developed to overcome the above-mentioned drawbacks of FIB-SEM. The scheme of a TXM setup is shown in Fig. 4.5 [85]. Full-field transmission X-ray microscopy operation principles are similar to that of the visible light transmission optical microscope, where the source of visible light is replaced by a bright X-ray generator. The 3D image is constructed by processing 2D projection data of a rotated sample. One advantage of TXM is that its non-deconstructive procedures can be maintained for relatively large measurement volumes without sacrificing imaging resolution, whereas FIB-SEM images of large volumes usually have poorer resolution. Another advantage is that TXM allows for accurate chemical phase segmentation as measurements can be done at X-ray energies above and below an elemental absorption edge, e.g. X-ray absorption

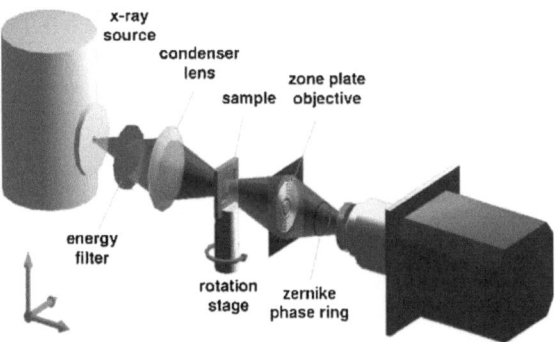

Fig. 4.5 TXM experimental setup showing an X-ray emission source with appropriate optics. The sample is mounted on a rotation stage, where hundreds of 2D projections are taken at a fixed exposure time per projection. A Zernike-phase ring is positioned in the back focal plane of the zone plate to enhance imaging contrast. Reprinted from Ref. [85]. Copyright (2008), with permission from the Electrochemical Society

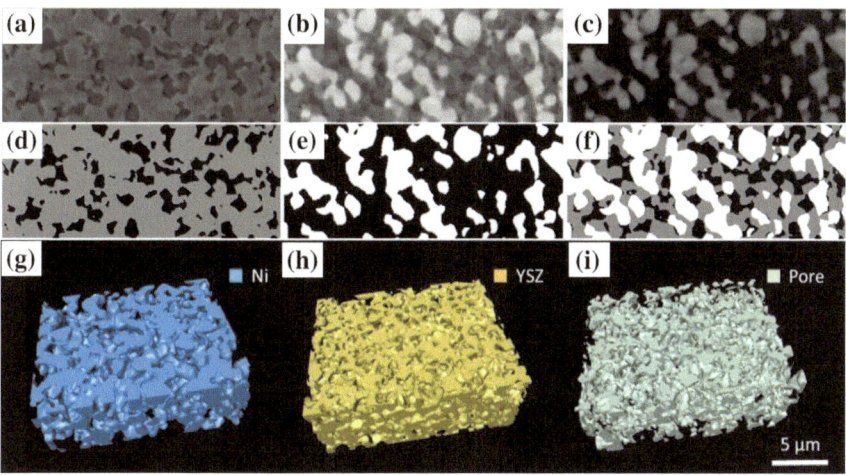

Fig. 4.6 TXM reconstructed slices of Ni-YSZ anode with X-ray **a** below and **b** above Ni absorption edge. **c** The result of subtracting below edge image from the above edge image, **d** segmented image of (**a**) to separate the solid phase (Ni + YSZ, *gray*) from the pore phase (*black*), **e** segmented image of (**c**) to isolate the Ni (*white*) and **f** the final result of segmentation showing Ni (*white*), YSZ (*gray*) and pore (*black*). **g–i** Segmented result of individual phases in one of the representative volumes. Reprinted from Ref. [86]. Copyright (2012), with permission from Elsevier

spectroscopy. An example about the phase segmentation of a Ni-YSZ anode is given in Fig. 4.6 [86]. However, high energy X-rays are required to avoid attenuation to penetrate bulk samples with large sizes [87]. Both high energy X-ray synchrotron sources and corresponding equipment setups are perforce to achieve high-resolution 3D images with large volume sizes.

The two aforementioned techniques, FIB-SEM and TXM, represent major advances in SOFC material characterization, in that, different phases, such as pores, Ni, (or LSM), and YSZ, can be visualized. They provide an accurate estimate of surface area, the porosity, tortuosity, as well as the location of all individual phases, and the triple-phase-boundary (TPB) length. Although both technologies require expensive and time-consuming focused ionic beam sharpening of SOFC samples, the future development of these techniques with higher resolution and larger imaging volume size will greatly facilitate the study of the correlation of microstructures of SOFC electrodes with the mass transport as well as electro-catalytic properties.

4.5 Correlations between Electrode Microstructures and SOFC Mass Transport

The relationship between electrode microstructures and SOFC mass transport has been studied either by experimental cell tests or by theoretical simulations. In experimental methods, the mass transport parameters, such as the gas effective diffusivity, the limiting current density, and the concentration polarization, can be directly obtained by fitting the cell's I–V curves. In this section, the effects of both anode and cathode microstructures on the mass transport are discussed based on the I–V curve fitting method. The influences of fuel and oxidant gas composition also are illustrated. Another experimental method, the electrochemical impedance spectra (EIS), is a powerful technique to separate quantitatively the mass transport information from the overall cell impedances. In addition to experimental methods, various models and numerical methods have been developed to investigate the gas diffusion in SOFCs. A brief introduction into these methods is given in the last part of this section.

4.5.1 I–V Curve Fitting

Based on the I-V fitting method as depicted in Sect. 3.3 in Chap. 3, Zhao and Virkar [88] investigated the effect of the anode thickness and porosity on SOFC performances. Their SOFCs was mounted in a test system as shown in Fig. 4.7a. The system facilitates the measurement of voltage versus current polarization under different temperatures. Figure 4.7b shows a SEM image of the cross section of a typical SOFC cell, which comprises a dense YSZ electrolyte and porous regions including a cathode interlayer, an anode interlayer, and an anode support structure.

The SOFC performance with respect to the Ni-YSZ anode thickness is shown in Fig. 4.8. The maximum power density (MPD) decreases from ~1.35 to ~0.7 W cm^{-2} as the anode support thickness increases from 0.5 mm to 2.45 μm, as shown in Fig. 4.8a. By fitting the voltage versus current density curve, Zhao et al. found that the effective binary diffusivity, $D_{H_2-H_2O}^{eff}$, varied little as the anode thickness increases. A more rapid decrease of cell voltage *versus* current density as the anode thickness increases arose from the larger ohmic resistance with increasing anode thickness. Further calculations showed that i_a decreased monotonous from 21.81 to 5.58 A cm^{-2} as the anode thickness increased in the range

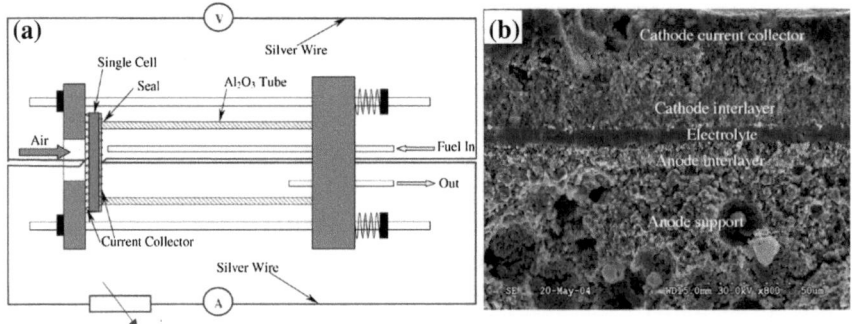

Fig. 4.7 a Schematic of the single cell testing apparatus. **b** An SEM micrograph of a typical cell. Reprinted from Ref. [88]. Copyright (2005), with permission from the Electrochemical Society

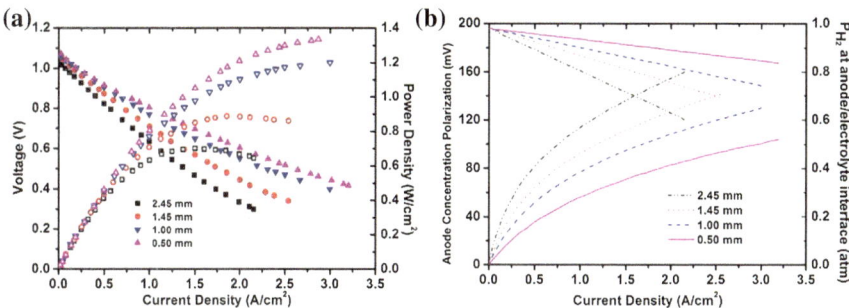

Fig. 4.8 **a** Voltage and power density versus current density plots at 800 °C for cells with different anode thicknesses. **b** Anode concentration polarization and partial pressure of hydrogen at the anode interlayer/electrolyte interface as a function of current density for cells with different anode support porosities. Reprinted from Ref. [88]. Copyright (2005), with permission from the Electrochemical Society

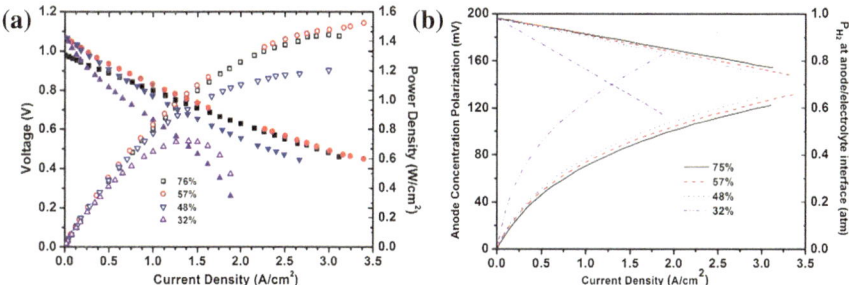

Fig. 4.9 **a** Voltage and power density versus current density for cells with anode porosity varied between 32 and 76 %. The OCV for the cell with 76 % anode support porosity is lower than the theoretical value, indicating that the YSZ electrolyte film was not gas-tight. **b** Anode concentration polarization, η_{anode}, and partial pressure of hydrogen at the anode interlayer/electrolyte interface as a function of current density for cells with different anode support porosities. Reprinted from Ref. [88]. Copyright (2005), with permission from the Electrochemical Society

of 0.5~2.45 μm. As shown in Fig. 4.8b, the anode concentration polarization increased dramatically with increasing the anode thickness. The result shows that making anodes as thin as possible is of significant to improve fuel gas transport and to decrease the energy loss by concentration polarization as long as the anode-supported SOFCs still maintain adequate mechanical strength.

The correlation of the anode support porosity with the SOFC performance also was studied by Zhao et al. Figure 4.9a shows that the MPD increased from ~0.7 to ~1.5 W cm^{-2} as the anode porosity increased from 32 to 57 %. A further increase in the anode porosity resulted in a slight decrease of MPD. Obtained by curve-fitting, the voltage versus current density curve shows that $D_{H_2-H_2O}^{eff}$ increased continuously from 0.22 to 0.82 cm^2 s^{-1} as the anode porosity increased from 32 to 76 %. The calculated anode concentration polarization by Eqs. 3.10 and 3.12

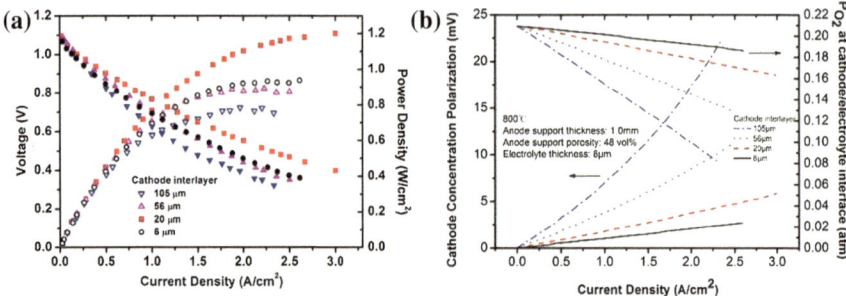

Fig. 4.10 a Voltage and power density versus current density for cells with cathode interlayer thickness varied between ~6 and ~105 μm. **b** Cathode concentration polarization, η_c, and the partial pressure of oxygen at cathode interlayer/electrolyte interface, $p^i_{O_2}$, as a function of cell current density for cells with different cathode interlayer thicknesses. Reprinted from Ref.[88]. Copyright (2005), with permission from the Electrochemical Society

shown in Fig. 4.9b indicated that η_{anode} increased continuously as the porosity increased since the contribution of electrochemical polarization is neglected in the calculation. The result above indicates that anodes with larger porosities in an appropriate range exhibit improved fuel gas transport and low ohmic resistance and, thus, result in low energy losses associated with SOFCs.

The correlation between cathode microstructures and SOFC mass transport was studied by Zhao et al. They investigated the effect of LSM cathode interlayer thickness on SOFC performance [88]. Figure 4.10a shows that the MPD decreased as the cathode interlayer thickness increased in the range of 20~105 μm. Yet, the cathode interlayer with a thinner thickness of 6 μm exhibited the lowest MPD. The effective binary diffusivity, $D^{eff}_{H_2-H_2O}$, estimated by fitting the voltage *versus* current density curve, varied little as the cathode interlayer thickness increases. According to the equation for η_c and i_c in Eqs. 3.11 and 3.13, i_c decreased and η_c increased as the cathode thickness increased. The result is verified by Fig. 4.10b showing that the cathode concentration increased dramatically with increasing the cathode thickness. Therefore, as long as an adequate, catalytically active triple-phase boundary is satisfied, thin electrodes should be employed to decrease SOFC energy losses caused by the concentration polarization.

Zhao et al. [89] have investigated the effect of cathode porosity on gas diffusion and concentration polarization. As shown in Fig. 4.11a, the effective binary diffusivity, $D^{eff}_{O_2-N_2}$ at 800 °C, increased from 0.016 to 0.12 cm²/s with increasing open porosity between 15 and 43 vol%. The slope of this increasing trend gew larger as the open porosity grew larger than 34 vol%, indicating the existence of a critical open porosity for SOFC cathodes. Figure 4.11b shows that the cathode overpotential increased with increasing current density. This increase became more gradual for cathodes with larger open porosity. The results indicate that cathodes with larger porosities show improved O_2 diffusion and much lower concentration polarization.

The effects of fuel composition on the gas transport and performance of SOFC also were investigated. Despite a very high conversion efficiency, a potentially

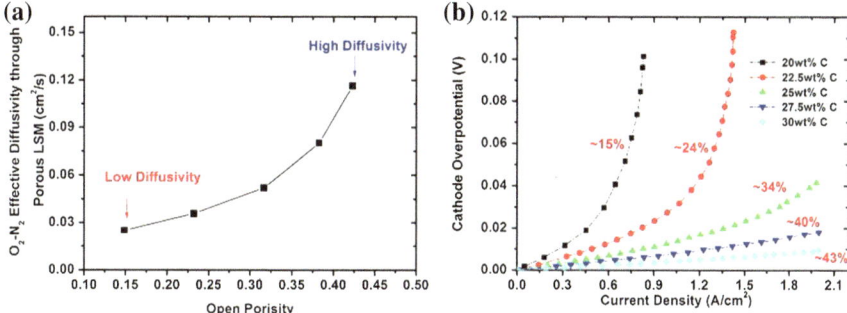

Fig. 4.11 **a** Effective binary diffusivity, $D^{eff}_{O_2-N_2}$ versus open porosity. **b** Calculated cathode overpotential as a function of current density for porous LSM cathodes fabricated with 20.0–30.0 wt% carbon. Cathode thickness = 200 mm. Reprinted from Ref. [89]. Copyright (2003), with permission from the Electrochemical Society

transformational benefit of SOFCs over low-temperature fuel cells is fuel flexibility. For instance, SOFCs can directly convert traditional fossil fuels into electricity without the restriction of Carnot cycling. Thus, thorough investigations of the effect of fuel composition to the performance of SOFCs are necessary. Jiang and Virkar [90] worked systematically and extensively on this issue. They compared the performance of SOFCs cells under different fuel and diluent mixtures conditions, including H_2–He, H_2–N_2, H_2–CO_2, H_2–H_2O, CO–CO_2, and H_2–CO. Graphs of the voltage versus current density and power density *versus* current density curves are shown in Fig. 4.12. Since the same SOFC cell with a specific Ni-YSZ anode, a YSZ-samaria-doped ceria (SDC) bilayer electrolyte, and a Sr-doped $LaCoO_3$ (LSC) + SDC cathode with specific microstructure is used for all the tests, the difference of the SOFC performance entirely came from fuel compositions. Concluding from Fig. 4.12, the cell performances increase as the fuel (H_2 or CO) concentration increases. For a given diluent concentration, the cell performance follows the order of H_2–CO > H_2–He > H_2–H_2O > H_2–N_2 > H_2–CO_2 > CO–CO_2. This result indicates that SOFC performance is better with an inert gas diluent of low molecular weight (such as He) than with an inert gas diluent of high molecular weight (such as N_2). H_2–CO is excluded in this molecular weight rule since CO is not an inert gas but an active fuel.

Jiang et al. further analyzed the curves of voltage versus current and obtained the limiting current density, i_a, with respect to the partial pressure of the fuel gas (either H_2 or CO). As shown in Fig. 4.13, i_a increased linearly as the partial pressure increases for all the fuel mixtures. H_2–He fuel (or H_2–H_2O–He mixtures) exhibits the highest i_{as} and the highest slope, and CO–CO_2 mixture exhibits the lowest i_{as} and slope. The binary diffusivity D_{12} can be calculated according to the Chapman–Enskog equation [91] and the effective ternary diffusivity by Eq. 4.1 derived from the Stefan–Maxwell equation [92],

$$D^{eff}_{H_2-mix} = \frac{i_a}{\dfrac{2Fp^0_{H_2}}{RTl_a} - \dfrac{i_aAp_t}{RTl_aJ}} \qquad (4.1)$$

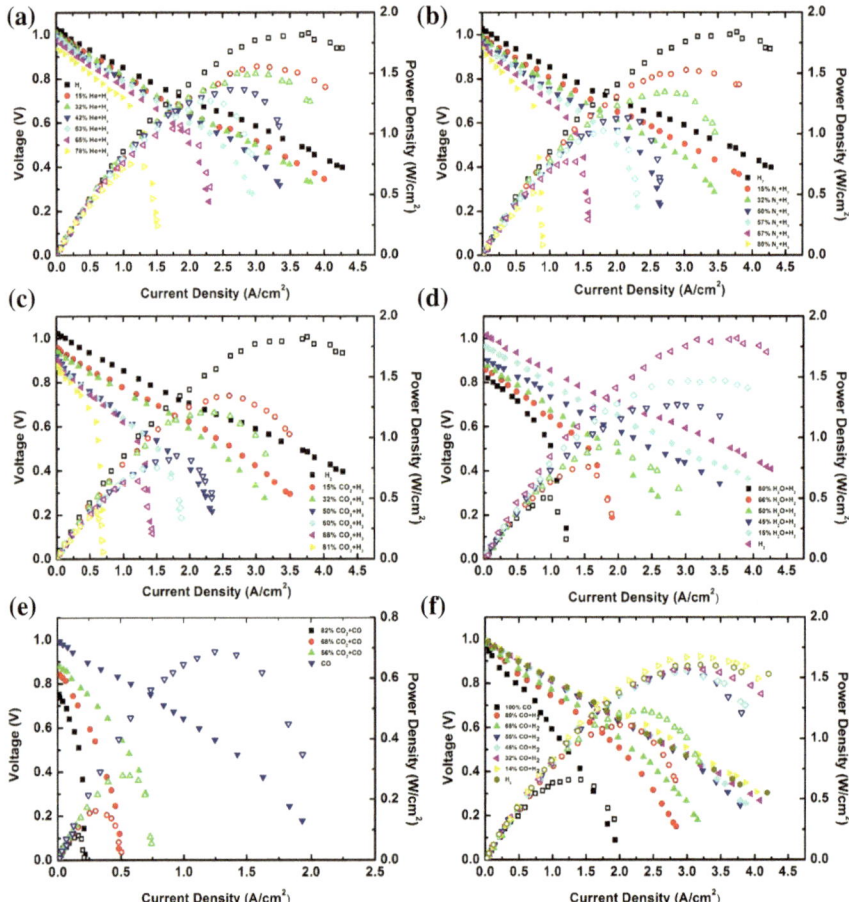

Fig. 4.12 Voltage and power density versus current density at 800 °C with **a** H_2–He mixture, **b** H_2–N_2 mixture, **c** H_2–CO_2 mixture, **d** H_2–H_2O mixture, **e** CO–CO_2 mixture, and **f** H_2–CO mixtures as fuel with various mixture ratios. Reprinted from Ref. [90]. Copyright (2003), with permission from the Electrochemical Society

where i_a is the anode-limiting current density, $p_{H_2}^0$ is the fuel partial hydrogen pressure at the anode, A is the electrode area, P_t is the total pressure, l_a is the anode thickness, and J is the total molar flow rate of the fuel and diluent. $D_{H_2-mix}^{eff}$ for H_2–H_2O is about five times larger than that for CO–CO_2, which indicates a lower cell performance with CO as a fuel compared with H_2 due to a higher concentration polarization with CO, regardless of the activity difference. Figure 4.13b shows the measured total polarization and the calculated concentration polarizations for H_2–H_2O and CO–CO_2, respectively. At a current density of 0.25 A cm^{-2}, the difference between the measured and the calculated concentration polarization

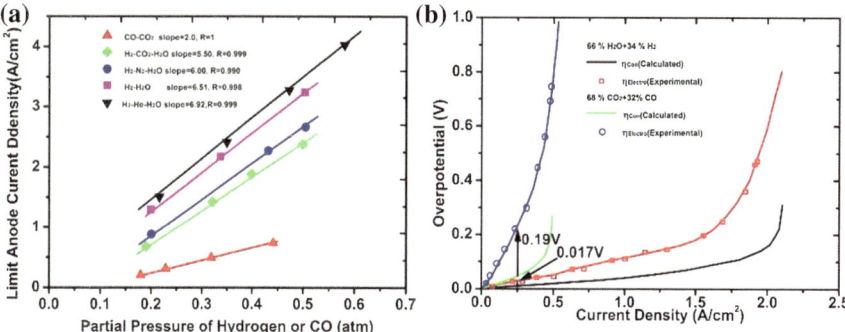

Fig. 4.13 **a** Anode-limiting current density versus partial pressure of H_2 or CO at 800 °C. **b** Comparison of the measured total polarization as a function of current density versus calculated concentration polarization for: ~34 % H_2 + 66 % H_2O and ~32 % CO + 68 % CO_2. Reprinted from Ref. [90]. Copyright (2003), with permission from the Electrochemical Society

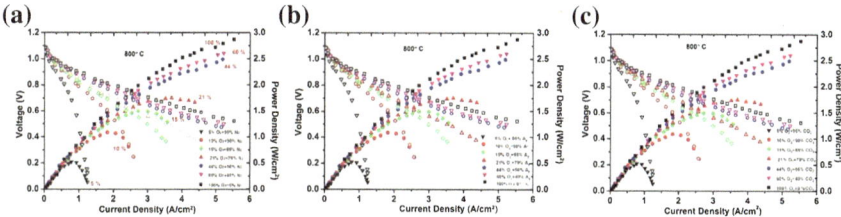

Fig. 4.14 Voltage and power density versus current density at 800 °C with different cathode oxidant compositions: **a** O_2–N_2, **b** O_2–Ar, **c** O_2–CO_2 mixtures with different ratios. Reprinted from Ref. [93]

is ~0.017 and 0.173 V, respectively, indicating the much larger anodic activation polarization for CO than that for H_2.

Virkar's group investigated the effect of cathode gas composition on the performance of SOFCs [93]. The same SOFC cell with a specific Ni-YSZ anode, YSZ-SDC bilayer electrolyte, and a Sr-doped $LaCoO_3$ (LSC) + SDC cathode with specific microstructure was employed as a test cell as cathode oxidant gases of different compositions were supplied. As shown in Fig. 4.14, the SOFC performance increases as the ratio of O_2 increases. A SOFC cathode with pure O_2 shows the highest power density of ~2.9 W cm^2 due to low cathode concentration polarization. At the same composition ratio, the SOFC performance follows the order O_2–N_2 > O_2–Ar > $O^{2-}CO_2$. The result indicates that cathode concentration polarization is lower with an inert diluent of low molecular weight (such as N_2) compared with an inert gas diluent of higher molecular weight (such as CO_2).

4.5.2 Electrochemical Impedance Spectra

Electrochemical impedance spectroscopy (EIS) is a powerful technique to investigate the effects of SOFC microstructural factors on their gas transport performance. The work done by Noh et al. [94] demonstrated the application of EIS technique in the gas transport study of the Ni-YSZ anodes. Three SOFCs with different anode supports are noted as T1, T2, and C-SP. The anode supports of T1 and T2 were made by the tape casting method by varying the amount of the plasticizer. C-SP was fabricated by screen printing. The other components of the three cells, including the anode interlayer, the dense YSZ electrolyte layer (1 μm), the GDC buffer layer, and ($La_{0.6}Sr_{0.4}CoO_{3-\delta}$) cathode were similar, as shown by cross-sectional SEM imaging in Fig. 4.15a–c. The porosity and pore size distribution of the anode supports were analyzed by a mercury porosimeter. The measured results demonstrate that the average pore size follows the order of T1 (127.4 nm) < T2 (195 nm) < C-SP (542 nm). Also, the porosity follows the same order of T1 (~18.7 %) < T2 (~23 %) < C-SP (~34 %). The gas transport performance of the three types of anode-supported cells was tested by EIS. When the partial pressure of the fuel at the anode side was varied, the low-frequency arc changed with respect to the fuel partial pressure (Fig. 4.15d), where the low-frequency arc corresponds to the mass transport of the anode. As shown in Fig. 4.15e, upon changing the cells from T1 to T2, it is noticeable that the low-frequency arc (frequency ~1 Hz)

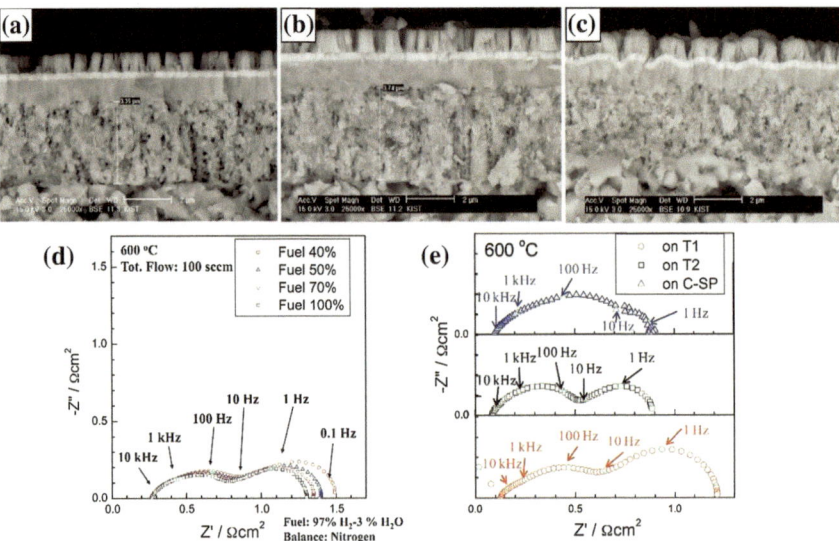

Fig. 4.15 Cross-sectional SEM micrographs of TF-SOFCs on **a** T1, **b** T2, and **c** C-SP anode supports. **d** Electrochemical impedance spectra of the same cell with the fuel partial pressure change. **e** Electrochemical impedance spectra of TF-SOFCs on T1, T2, and C-SP anode supports measured at open-circuit potential. Reprinted from Ref. [94]. Copyright (2011), with permission from Elsevier

notably reduced. Upon the change from tape anodes (T1 and T2) to C-SP, the low-frequency arc of the TF-SOFC on C-SP was remarkably smaller than those of the TF-SOFC on tape anodes. Therefore, the EIS data indicate that the fuel gas transport is improved as the pore size and the porosity of the anodes increase.

Despite qualitative analysis of the gas diffusion impedance separation from the overall electrode process, the EIS also has been developed to ascertain quantitatively gas diffusion impedances of SOFC electrodes. For instance, Hussain et al. [95] evaluated the gas diffusion impedance of a symmetric cell, fabricated by screen printing Nb-doped $SrTO_3$ (STN, a type of anode materials used in low-temperature SOFCs) on both sides of a dense ScYSZ electrolyte (10 mol% Sc_2O_3 and 1 mol% Y_2O_3 stabilized ZrO_2). STN anodes with different microstructures (porosity, and pore size, etc.) were prepared by infiltrating Ni-GDC into the porous STN backbones. By varying the number of times that the microstructures were infiltrated (from 1× to 9×), the loaded amount of Ni-GDC could be tuned. Figure 4.16 shows the SEM images of two infiltrated STN backbones with the same number of infiltrations. Backbone-I as shown in Fig. 4.16a, b has a larger particle size and smaller porosity compared with that of backbone-II (Fig. 4.16a, b). Figure 4.17 shows the impedance spectra of the STN backbone-I and backbone-II infiltrated with Ni-GDC electro-catalyst. An equivalent circuit containing the electrolyte series resistance, R_0, in series with a R_1Q_1 (R_1 is the electrode process resistance, and Q is a constant phase element), which are in parallel with each other, all of which are in series with

Fig. 4.16 SEM microstructure of Ni-GDC (5×) infiltrated STN anodes (after electrochemical testing) **a** backbone-I, **b** backbone-I with higher magnification, **c** backbone-II and **d** backbone-II with higher magnifications. Reprinted from Ref. [95]. Copyright (2012), with permission from International Association of Hydrogen Energy

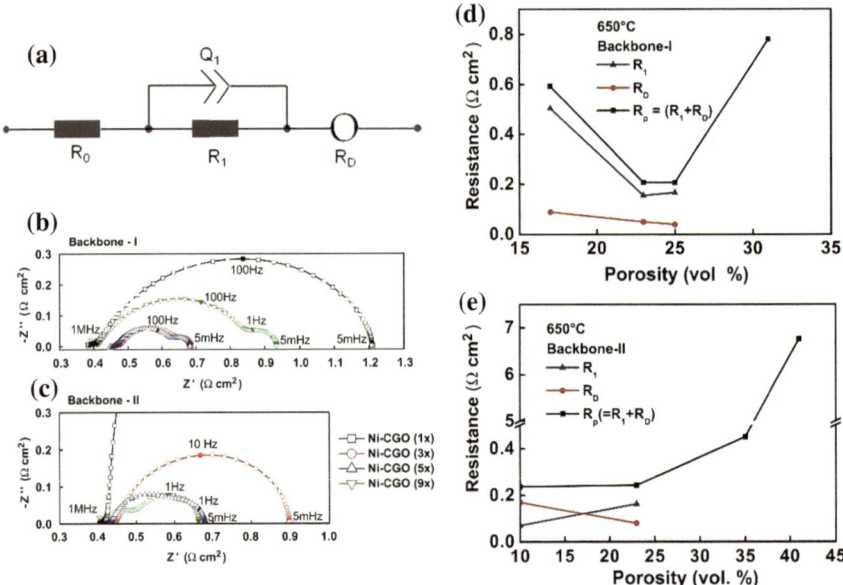

Fig. 4.17 a Schematic representation of an equivalent circuit $R_0(R_1Q_1)(O)$. **b, c** Impedance spectra at 650 °C in $H_2/3$ % H_2O for Ni-GDC ($1\times$, $3\times$, $5\times$, and $9\times$) infiltrated on STN **b** backbone-I and **c** backbone-II. **d, e** Rp versus porosity of Ni-GDC-infiltrated STN anodes at 650 °C for **d** backbone-I and **e** backbone-II. Reprinted from Ref. [95]. Copyright (2012), with permission from International Association of Hydrogen Energy

a finite-length Warburg element (O) in series, i.e., $R_0(R_1Q_1)(O)$, was used to model the spectra as shown in Fig. 4.17a. The finite-length Warburg element determines the value of gas diffusion resistance, R_D. R_D can be calculated by the complex impedance ($Y(\omega)$) and the admittance parameter (Y_0, in $S{\cdot}s^{1/2}$), given by Eqs. 4.2–4.4 [96, 97],

$$R_D = \frac{B}{Y_0} \tag{4.2}$$

$$Y(\omega) = \frac{\tanh\left[B \cdot \sqrt{(j \cdot \omega)}\right]}{Y_0 \cdot \sqrt{(j \cdot \omega)}} \tag{4.3}$$

$$Y_0 = \frac{n \cdot F \cdot s\sqrt{D_{A-mix}}}{V_m} \left|\frac{dE}{dy}\right|^{-1} \tag{4.4}$$

where s is the interface area, V_m is the molar volume, and $|dE/dy|$ is the slope of the electrochemical titration curve relating voltage and concentration. B is the time constant, in $s^{1/2}$. B and Y_0 were obtained from the equivalent circuit fittings. The effective ternary diffusion coefficient of gas A, D_{A-mix}, is determined from Eq. 4.5 [98],

$$D_{A-\text{mix}} = \frac{1 - X_A}{\frac{X_B}{D_{AB}} + \frac{X_C}{D_{AC}} + \cdots} \tag{4.5}$$

where X_A, X_B, X_C are the molar fractions in the ternary gas mixture. D_{AB} and D_{AC} correspond to the respective effective binary diffusion coefficients. Figure 4.17b, c present the impedance spectra of the STN backbone-I and backbone-II by varying the infiltration time of Ni-GDC at 650 °C in H_2/3 % H_2O. To analyze the impedance spectra of the infiltrated electrodes in detail, the effect of porosity in the electrode polarization resistances for backbone-I and backbone-II at 650 °C was plotted in Fig. 4.17d, e. This assessment separates the total polarization cell resistance, R_p into an electrode activation resistance, R_1, and gas diffusion resistance, R_D. R_1 reaches a minimum at a critical porosity of 23% for the infiltrated anodes due to the increased loading amount of electro-catalyst. Thereafter, R_1 increases as a function of porosity due to the overloading of electro-catalyst. The increase of R_D with increasing porosity in both backbones indicates that Knudsen diffusion in the porous electrode plays a role in the total gas diffusion impedances. Therefore, a moderate infiltrating amount of electro-catalyst should be employed in the fabrication of an SOFC electrodes with both low activation resistance and gas diffusion resistance. Other researchers have contributed to the quantitative assessment of the gas diffusion resistance using EIS techniques by building various mathematical models and equivalent circuits [96–106]. More reasonable and general models are still needed to make EIS analysis more accurate and reliable in the field of mass transport within SOFCs.

4.5.3 Theoretical Simulations

Various numeric models, from simple mathematical models to finite-element method and lattice Boltzmann methods, were built to investigate correlations associated with mass transport performance of SOFCs with electrode microstructures [97–116]. These models for mass transport inside porous SOFC electrodes were developed based on the gas diffusion models discussed in Chap. 2, including Fick's model, the dusty-gas model (DGM) and the Stefan–Maxwell model (SMM), to predict the concentration overpotential. For instance, Li et al. [117] employed a finite-element method to investigate a cell's performance by using a theoretical model that considered heterogeneous elementary reactions, electrochemical reactions, electrode microstructure, and mass and charge transport. For mass transport, molecular diffusion and Knudsen diffusion are considered in the porous electrodes. Convection diffusion caused by a pressure gradient and surface diffusion of microscale surface species were ignored. The effects of microstructure, thickness, and temperature of cathodes on the cell performance in both solid oxides are discussed. The mean particle diameters of electrodes are assumed to be the same as the mean pore diameter. The results indicate that the increasing thickness of the cathode induces an increase of the cell performance first and then a decrease as a consequence of the combined effects of the

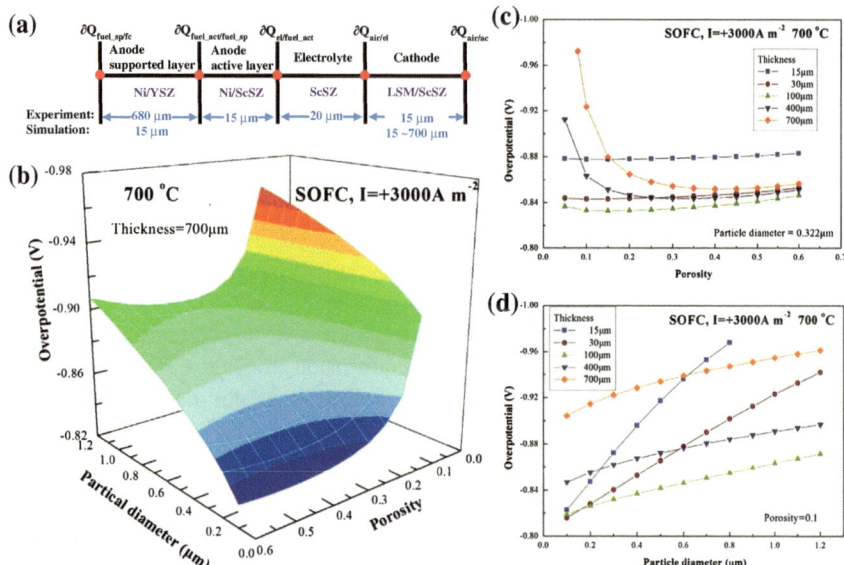

Fig. 4.18 **a** Simplified modeling geometry. **b** Effects of cathode particle diameter and porosity on overpotential in a surface plot. *Curve* plots in **c**, **d** shows the effect of **c** the porosity and **d** particle size on cell overpotential for cathodes of different thicknesses. Reprinted from Ref. [117]. Copyright (2014), with permission from International Association of Hydrogen Energy

more reaction active sites and the larger gas diffusion resistance. Therefore, an optimal thickness should exist. As the porosity increases, the cell performance first gradually decreases due to the gas diffusion limitation and then slightly increases, due to the decreasing reaction active sites. An optimized porosity should also be employed. Reducing the particle diameter can significantly increase the performance, which is mainly attributed to the square growth rate of reaction active sites (Fig. 4.18).

Although numerous methods and models have been developed to investigate the mass transport in SOFCs, 3D structural information regarding both electrode and electrolyte remains lacking in these simulations, resulting in deviations from experimental values. As structural characterization advances, especially the rapid development of the 3D imaging techniques, like FIB-SEM and TXM, 3D information will become more readily available. Concurrently, more accurate simulations are expected based on these data through the implementation of advanced mass diffusion models that are applicable to both micro- and nanostructured electrodes and electrolytes.

4.6 Summary

In conclusion, the gas transport characteristics of both anodes and cathodes in SOFCs were reviewed, with the aim to optimize performance by minimizing the concentration polarization. We started our depiction by emphasizing high catalytic electrode materials and the control of their microstructure. 3D imaging techniques were reviewed that are involved in the characterization of the microstructures of SOFC electrodes. Factors related to gas transport, obtained quantitatively through these techniques, including porosity, pore size, surface area, and tortuosity, were assessed. The correlation between the electrode microstructure and the gas diffusion performance was discussed, guided by recent research progress and insight. Future research trajectories should focus on facile methods to fabricate both anode and cathode materials with desired porosity, pore size, and small tortuosity as well as long TPB length of high activity. Existing 3D imaging techniques, such as FIB-SEM and TXM, should be further improved to acquire even higher resolution images of large volumes in a cost-effective and time efficient fashion. The accurate evaluation of the diffusion-related factors for newly developed, high-active anode and cathode materials of SOFCs remains inadequate; vigorous investigation is thus needed.

Undoubtedly, gas transport performance is an important aspect in the research of SOFCs. Much work in theoretical simulations, experimental characterization, and cell test has been done to understand the relationship between the electrode microstructures and the concentration polarization. Further studies on this subject along with steady progress on the development of new electrode materials and architectures, novel nondestructive 3D image techniques, and long-term stability of the electrode gas transport performance should be pursued to enhance dramatically the energy conversion efficiency of SOFCs.

References

1. E.D. Wachsman, K.T. Lee, Lowering the temperature of solid oxide fuel cells. Science **334**, 935–939 (2011)
2. J.C. Ruiz-Morales, J. Canales-Vazquez, C. Savaniu et al., Nature **439**, 568–571 (2006)
3. Y.H. Huang, R.I. Dass, Z.L. Xing et al., Science **312**, 254–257 (2006)
4. Y.B. Kim, T.P. Holme, T.M. Gür, F.B. Prinz, Adv. Funct. Mater. **21**, 4684–4690 (2011)
5. C.S. Song, Fuel processing for low-temperature and high-temperature fuel cells challenges, and opportunities for sustainable development in the 21st century. Catal. Today **77**, 17 (2002)
6. B.C.H. Steele, A. Heinzel, Materials for fuel-cell technologies. Nature **414**, 345 (2001)
7. Y. Arachi, H. Sakai, O. Yamamoto et al., Electric conductivity of ZrO_2-Ln_2O_3 (Ln = lanthanides) system. Solid State Ionics **121**(1–4), 133–139 (1999)
8. S.P. Jiang, J. Mater. Sci. **43**, 6799–6833 (2008)
9. B.C.H. Steele, Solid State Ionics **75**, 157 (1995)
10. D.W. Jung, L.L. Duncan, E.D. Wachsman, Acta Mater. **58**, 355 (2010)
11. D. Andersson, S. Simak, N. Skorodumova et al., Optimization of ionic conductivity in doped ceria, PNAS **103**(10), 3518–3521 (2006)
12. X.M. Ge, S.H. Chan, Q.L. Liu, Q. Sun, Adv. Energy Mater. **2**, 1156 (2012)
13. S. Park, J. Vohs, R. Gorte, Direct oxidation of hydrocarbons in a solid-oxide fuel cell. Nature **404**, 265–267 (2000)
14. W.Z. Zhu, S.C. Deevi, A review on the status of anode materials for solid oxide fuel cells. Mater. Sci. Eng. **A362**, 228–239 (2003)
15. Xinwen Zhou, Ning Yan, Karl T. Chuang, Jingli Luo, Progress in La-doped $SrTiO_3$ (LST)-based anode materials for solid oxide fuel cells. RSC Adv. **4**, 118 (2014)

16. A. Atkinson, S. Barnett, R.J. Gorte, J.T.S. Irvine, A.J. Mcevoy, M. Mogensen, S.C. Singhal, J. Vohs, Nat. Mater. **3**, 17–27 (2004)

17. P.I. Cowin, C.T.G. Petit, R. Lan, J.T.S. Irvine, S. Tao, Adv. Energy Mater. **1**, 314 (2011)

18. B. Shri Prakash, S. Senthil Kumar, S.T. Aruna, Properties and development of Ni/YSZ as an anode material in solid oxide fuel cell: a review. Renew. Sustain. Energy Rev. **36**, 149–179 (2014)

19. L. Tai, M. Nassrallah, H. Anderson et al., Structure and electrical-properties of $La_{1-x}Sr_xCo_{1-y}Fe_yO_3$. The system of $La_{0.8}Sr_{0.2}Co_{1-y}Fe_yO_3$. Solid State Ionics **76**(3–4), 259–271 (1995)

20. H.Y. Tu, Y. Takeda, N. Imanishi et al., $Ln_{0.4}Sr_{0.6}Co_{0.8}Fe_{0.2}O_3$ (Ln = La, Pr, Nd, Sm, Gd) for the electrode in solid oxide fuel cells. Solid State Ionics **117**, 277–281 (1999)

21. Z. Shao, S.M Haile, a high-performance cathode for the next generation of solid-oxide fuel cells. Nature **431**(7005), 170–173 (2004)

22. W. Zhou, R. Ran, Z. Shao, J. Power Sources **192**, 231–246 (2009)

23. H.Y. Tu, Y. Takeda, N. Imanishi et al., $Ln_{1-x}Sr_xCoO_3$ (Ln = Sm, Dy) for the electrode of solid fuel cells. Solid State Ionics **100**(3), 283–288 (1997)

24. D. Ding, X. Li, S.Y. Lai, K. Gerdes, M. Liu, Enhancing SOFC cathode performance by surface modification through infiltration. Energy Environ. Sci. **7**, 552–575 (2014)

25. S.J. Skinner, Recent advances in Perovskite-type materials for solid oxide fuel cell cathodes. Int. J. Inorg. Mater. **3**, 113–121 (2001)

26. A.J. Jacobson, Materials for solid oxide fuel cells. Chem. Mater. **22**, 660–674 (2010)

27. C. Sun, R. Hui, J. Roller, Cathode materials for solid oxide fuel cells: a review. J. Solid State Electrochem. **14**, 1125–1144 (2010)

28. D.H. Dong, M.F. Liu, K. Xie, J. Sheng, Y.H. Wang, X.B. Peng, X.Q. Liu, G.Y. Meng, J. Power Sources **175**, 201 (2008)

29. R. Campana, A. Larrea, J.I. Pen, V.M. Orera, J. Eur. Ceram. Soc. **29**, 85 (2009)

30. T. Suzuki, Y. Funahashi, T. Yamaguchi, Y. Fujishiro, M. Awano, J. Power Sources **183**, 544 (2008)

31. M.F. Liu, J.F. Gao, G.Y. Meng, Comparative study on the performance of tubular and button cells with YSZ membrane fabricated by a refined particle suspension coating technique. Int. J. Hydrogen Energy **35**, 10489–10494 (2010)

32. J.E. Hong, T. Inagaki, T. Ishihara, Preparation of LaGaO3 thin film for intermediate temperature SOFC by screen-printing method (I). Ionics **18**, 433–439 (2012)

33. J. Wang, Z. Lu, K. Chen, Study of slurry spin coating technique parameters for the fabrization of anode-supported YSZ films for SOFCs. J. Power Source **164**, 17–23 (2007)

34. Weidong He, K.J. Yoon, R.S. Eriksen, S. Gopalan, S.N. Basu, U.B. Pal, J. Power Sources **195**, 532–535 (2010)

35. K.J. Yoon, S. Gopalan, U.B. Pal, J. Electrochem. Soc. **156**(3), B311–B317 (2009)

36. A.C. Muller, D. Herbstritt, E. Ivers-Tiffee, Development of a multilayer anode for solid oxide fuel cells. Solid State Ionics **152**, 537–542 (2002)

37. P. Holtappels, C. Sorof, M.C. Verbraeken, S. Rambert, U. Vogt, Preparation of porosity-graded SOFC anode substrates. Fuel Cells **6**, 113–116 (2006)

38. S.W. Sofie, Fabrication of functionally graded and aligned porosity in thin ceramic substrates with the novel freeze–tape-casting process. J. Am. Ceram. Soc. **90**, 2024–2031 (2007)

39. Y. Chen, J. Bunch, T.S. Li, Z.P. Mao, F.L. Chen, Novel functionally graded acicular electrode for solid oxide cells fabricated by the freeze-tape-casting process. J. Power Sources **213**, 93–99 (2012)

40. A. Chung Min, S. Jung-Hoon, K. Inyong, Sammes N., The effect of porosity gradient in a Nickel/Yttria stabilized Zirconia anode for an anode-supported planar solid oxide fuel cell. J. Power Sources **195**, 821–824 (2010)

41. L. Chen, M. Yao, C. Xia, Anode substrate with continuous porosity gradient for tubular solid oxide fuel cells. Electrochem. Commun. **38**, 114–116 (2014)

42. D.H. Dong, J.F. Gao, X.Q. Liu, G.Y. Meng, Fabrication of tubular NiO/YSZ anode-support of solid oxide fuel cell by gelcasting. J. Power Sources **165**, 217–223 (2007)

43. D. Dong, X. Shao, K. Xie, X. Hua, G. Parkinson, C.Z. Li, Microchanneled anode supports of solid oxide fuel cells. Electrochem. Commun. **42**, 64–67 (2014)
44. C. Jin, C.H. Yang, F.L. Chen, Effects on microstructure of NiO–YSZ anode support fabricated by phase-inversion method. J. Membr. Sci. **363**, 250–255 (2010)
45. X. Shao, D.H. Dong, G. Parkinson, C.Z. Li, Microstructure control of oxygen permeation membranes with templated microchannels. J. Mater. Chem. A **2**, 410–417 (2014)
46. X. Shao, D. Dong, G. Parkinson, C.Z. Li, A microchanneled ceramic membrane for highly efficient oxygen separation. J. Mater. Chem. A **1**, 9641–9644 (2013)
47. B.F.K. Kingsbury, K. Li, A morphological study of ceramic hollow fibre membranes. J. Membr. Sci. **328**, 134–140 (2009)
48. M.H.D. Othman, Z.T. Wu, N. Droushiotis, G. Kelsall, K. Li, Morphological studies of macrostructure of Ni-CGO anode hollow fibres for intermediate temperature solid oxide fuel cells. J. Membr. Sci. **360**, 410–417 (2010)
49. N. Droushiotis, M.H.D. Othman, U. Doraswami, Z.T. Wu, G. Kelsall, K. Li, Novel co-extruded electrolyte–anode hollow fibres for solid oxide fuel cells. Electrochem. Commun. **11**, 1799 (2009)
50. C.H. Yang, C. Jin, F.L. Chen, Micro-tubular solid oxide fuel cells fabricated by phase-inversion method. Electrochem. Commun. **12**, 657–660 (2010)
51. W.N. Yin, B. Meng, X.X. Meng, X.Y. Tan, Highly asymmetric yttria stabilized zirconia hollow fibre membranes. J. Alloys Compd. **476**, 566 (2009)
52. S. Peng, Y. Wei, J. Xue, Y. Chen, H. Wang, Pr1.8La0.2Ni0.74Cu0.21Ga0.05O4Dd as a potential cathode material with CO_2 resistance for intermediate temperature solid oxide fuel cell. Int. J. Hydrogen Energy **38**, 10552 (2013)
53. C. Yang, C. Jin, M. Liu, F. Chen, Intermediate temperature micro-tubular SOFCs with enhanced performance and thermal stability. Electrochem. Commun. **34**, 231 (2013)
54. W. Sun, N.Q. Zhang, Y.C. Mao, K.N. Sun, Fabrication of anode-supported Sc_2O_3-stabilized-ZrO_2 electrolyte micro-tubular solid oxide fuel cell by phase-inversion and dip-coating. Electrochem. Commun. **20**, 117 (2012)
55. H. Wang, J. Liu, Effect of anode structure on performance of cone-shaped solid oxide fuel cells fabricated by phase inversion. Int. J. Hydrogen Energy **37**, 4339 (2012)
56. M.H.D. Othman, N. Droushiotis, Z. Wu, G. Kelsall, K. Li, High-performance, anode-supported, microtubular SOFC prepared from single-step-fabricated, dual-layer hollow fibers. Adv. Mater. **23**, 2480 (2011)
57. M.H.D. Othman, N. Droushiotis, Z.T. Wu, G. Kelsall, K. Li, Novel fabrication technique of hollow fibre support for micro-tubular solid oxide fuel cells. J. Power Sources **196**, 5035 (2011)
58. C. Yang, W. Li, S. Zhang, L. Bi, R. Peng, C. Chen, W. Liu, Fabrication and characterization of an anode-supported hollow fiber SOFC. J. Power Sources **187**, 90 (2009)
59. C. Jin, J. Liu, L.H. Li, Y.H. Bai, Electrochemical properties analysis of tubular NiO-YSZ anode-supported SOFCs fabricated by the phase-inversion method. J. Membr. Sci. **341**, 233–237 (2009)
60. C.C. Chen, M.F. Liu, M.L. Liu, Anode-supported micro-tubular SOFCs fabricated by a phase-inversion and dip-coating process. Int. J. Hydrogen Energy **36**, 5604–5610 (2011)
61. S. Brunauer, P.H. Emmett, E. Teller, J. Am. Chem. Soc. **60**(309), 10 (1938)
62. L. Holzer, F. Indutnyi, P.H. Gasser, B. Munch, M. Wegmann, J. Microsc. **216**, 84 (2004)
63. T. Suzuki, Z. Hasan, Y. Funahashi, T. Yamaguchi, Y. Fujishiro, M. Awano, Impact of anode microstructure on solid oxide fuel cells. Science **325**, 852–855 (2009)
64. S.B. Adler, Factors governing oxygen reduction in solid oxide fuel cell cathodes. Chem. Rev. **104**, 4791–4843 (2004)
65. Z. Shao, W. Zhou, Z. Zhu, Advanced synthesis of materials for intermediate-temperature solid oxide fuel cells. Prog. Mater. Sci. **57**, 804–874 (2012)
66. J.R. Wilson, W. Kobsiriphat, R. Mendoza, H.-Y. Chen, J.M. Hiller, D.J. Miller et al., Nat. Mater. **5**, 541–544 (2006)

67. L.A. Giannuzzi, F.A. Stevie, *Introduction to focused ion beams: instrumentation, theory techniques and practice* (Springer, Berlin, 2005)
68. M. Kishimoto, H. Iwai, M. Saito, H. Yoshida, Quantitative evaluation of SOFC porous anode microstructure based on FIB-SEM technique and prediction of anode overpotentials. J. Power Sources **196**, 4555–4563 (2011)
69. G.J. Nelson, W.M. Harris, J.J. Lombardo, J.R. Izzo, W.K.S. Chiu, P. Tanasini et al., Electrochem. Commun. **13**, 586–589 (2011)
70. J.R. Wilson, J.S. Cronin, S.A. Barnett, Scripta Mater. **65**, 67–72 (2011)
71. J.S. Cronin, J.R. Wilson, S.A. Barnett, J. Power Sources **196**, 2640–2643 (2011)
72. L. Holzer, B. Munch, B. Iwanschitz, M. Cantoni, T. Hocker, T. Graule, J. Power Sources **196**, 7076–7089 (2011)
73. H. Iwai, N. Shikazono, T. Matsui, H. Teshima, M. Kishimoto, R. Kishida et al., J. Power Sources **195**, 955–961 (2009)
74. Y. Karen Chen-Wiegart, R. DeMike, C. Erdonmez, K. Thornton, S.A. Barnett, J. Wang, J. Power Sources **249**, 349–356 (2014)
75. K. Yakal-Kremski, J.S. Cronin, Y.-C.K. Chen-Wiegart, J. Wang, S.A. Barnett, Fuel Cells **13**(4), 449–454 (2013)
76. Z. Jiao, N. Shikazono, N. Kasagi, J. Electrochem. Soc. **159**(3), B285–B291 (2012)
77. Y. Guan, Y.H. Gong, W.J. Li, J. Gelb, L. Zhang, G. Liu, X.B. Zhang, X. Song, C. Xia, Y. Xiong, H.Q. Wang, H. Wang, W.B. Yun, Y.C. Tian, J. Power Sources **196**, 10601–10605 (2011)
78. P.R. Shearing, R.S. Bradley, J. Gelb, S.N. Lee, A. Atkinson, P.J. Withers et al., Electrochem. Solid State Lett. **14**, B117–B120 (2011)
79. P.R. Shearing, J. Gelb, J. Yi, W.K. Lee, M. Drakopolous, N.P. Brandon, Electrochem. Commun. **12**, 1021–1024 (2010)
80. K.N. Grew, Y.S. Chu, J. Yi, A.A. Peracchio, J.R. Izzo, Y. Hwu, F. De Carlo, W.K.S. Chiu, J. Electrochem. Soc. **157**(6), B783–B792 (2010)
81. G.J. Nelson, J.R. Izzo, J.J. Lombardo, W.M. Harris, A.P. Cocco, W.K.S. Chiu, K.N. Grew, A. Faes, A. Hessler-Wyser, J. Van Herle, Y.S. Chu, S. Wang, ECS Trans. **35**(1), 1323–1327 (2011)
82. G.J. Nelson, K.N. Grew, J.R. Izzo, J.J. Lombardo, W.M. Harris, A. Faes, A. Hessler-Wyser, J. Van Herle, S. Wang, Y.S. Chu, A.V. Virkar, W.K.S. Chiu, Acta Mater. **60**, 3491–3500 (2012)
83. J. Laurencin, R. Quey, G. Delette, H. Suhonen, P. Cloetens, P. Bleuet, J. Power Sources **198**, 182–189 (2012)
84. P.R. Shearing, J. Golbert, R.J. Chater, N.P. Brandon, Chem. Eng. Sci. **64**, 3928–3933 (2009)
85. J.R. Izzo, A.S. Joshi, K.N. Grew, W.K.S. Chiu, A. Tkachuk, S.H. Wang, W. Yunb, Nondestructive reconstruction and analysis of SOFC anodes using X-ray computed tomography at sub-50 nm resolution. J. Electrochem. Soc. **155**(5), B504–B508 (2008)
86. Y. Karen Chen-Wiegart, J.S. Cronin, Q. Yuan, K.J. Yakal-Kremski, S.A. Barnett, J. Wang, 3D Non-destructive morphological analysis of a solid oxide fuel cell anode using full-field X-ray nano-tomography. J. Power Sources **218**, 348–351 (2012)
87. J. Gelb, W.B. Yun, M. Feser, A. Tkachuk, D.G. Seiler, A.C. Diebold, R. McDonald, C.M. Garner, D. Herr, R.P. Khosla, E.M. Secula, Frontiers of characterization and metrology for nanoelectronics: X-ray microscopy for interconnect characterization. AIP Conf. Proc. **1173**, 145–148 (2009)
88. F. Zhao, A.V. Virkar, J. Power Sources **141**, 79–95 (2005)
89. F. Zhao, T.J. Armstrong, A.V. Virkar, J. Electrochem. Soc. **150**(3), A249–A256 (2003)
90. Yi Jiang, Anil V. Virkar, J. Electrochem. Soc. **150**(7), A942–A951 (2003)
91. E.L. Cussler, *Diffusion: mass transfer in fluid systems* (Cambridge University Press, Cambridge, 1995)
92. R. Jackson, *Transport in porous catalysts* (Elsevier, Amsterdam, 1977)
93. A.V. Virkar, Y. Jiang, T.J. Armstrong, F. Zhao, N. Tikekar, S. Shinde, *Research on SOFC electrodes* (SECA Workshop, Pittsburgh, 2002), pp. 18–19

94. H.S. Noh, H. Lee, B.K. Kim, H.W. Lee, J.H. Lee, J.W. Son, Microstructural factors of electrodes affecting the performance of anode-supported thin film yttria-stabilized zirconia electrolyte (~1 μm) solid oxide fuel cells. J. Power Sources **196**, 7169–7174 (2011)

95. A. Mohammed Hussain, J.V.T. Høgh, T. Jacobsen, N. Bonanos, Nickel-ceria infiltrated Nb-doped $SrTiO_3$ for low temperature SOFC anodes and analysis on gas diffusion impedance. Int. J. Hydrogen Energy **37**(5), 4309–4318 (2012)

96. B.A. Boukamp, A nonlinear least squares fit procedure for analysis of immittance data of electrochemical systems. Solid State Ionics **20**, 31–44 (1986)

97. I.D. Raistrick, R.A. Huggins, The transient electrical response of electrochemical systems containing insertion reaction electrodes. Solid State Ionics **7**, 213–218 (1982)

98. D.F. Fairbanks, C.R. Wilke, Diffusion coefficients in multicomponent gas mixtures. Ind. Eng. Chem. **42**, 471–475 (1950)

99. A. Leonidea, S. Hansmanna, A. Webera, E. Ivers-Tiffée, Performance simulation of current/voltage-characteristics for SOFC single cell by means of detailed impedance analysis. J. Power Sources **196**, 7343–7346 (2011)

100. Q.A. Huang, R. Hui, B. Wang, J. Zhang, A review of AC impedance modeling and validation in SOFC diagnosis. Electrochim. Acta **52**, 8144–8164 (2007)

101. S.H. Jensen, A. Hauch, P.V. Hendriksen, M. Mogensen, N. Bonanos, T. Jacobsenb, A method to separate process contributions in impedance spectra by variation of test conditions. J. Electrochem. Soc. **154**(12), B1325–B1330 (2007)

102. A. Leonidea, Y. Apela, E. Ivers-Tifféea, SOFC modeling and parameter identification by means of impedance spectroscopy. ECS Trans. **19**(20), 81–109 (2009)

103. R.U. Payne, Y. Zhu, W.H. Zhu, M.S. Timper, S. Elangovan, B.J. Tatarchuk, Diffusion and gas conversion analysis of solid oxide fuel cells at loads via AC impedance. Int. J. Electrochem. **2011**, 1–11 (2011)

104. T. Jacobsen, P.V. Hendriksen, S. Koch, Diffusion and conversion impedance in solid oxide fuel cells. Electrochim. Acta **53**, 7500–7508 (2008)

105. S.P. Yoona, S.W. Nama, J. Hana, T.H. Lima, S.A. Honga, S.H. Hyun, Effect of electrode microstructure on gas-phase diffusion in solid oxide fuel cells. Solid State Ionics **166**, 1–11 (2004)

106. S. Primdahl, M. Mogensen, Gas diffusion impedance in characterization of solid oxide fuel cell anodes. J. Electrochem. Soc. **146**(8), 2827–2833 (1999)

107. H. Xu, Z. Dang, B-F. Bai, Numerical simulation of multispecies mass transfer in a SOFC electrodes layer using Lattice Boltzmann method. J. Fuel Cell Sci. Technol. **9**, 061004-1-6 (2012).

108. S. Kakac, A. Pramuanjaroenkij, X.Y. Zhou, A review of numerical modeling of solid oxide fuel cells. Int. J. Hydrogen Energy **32**(7), 761–786 (2007)

109. M. Andersson, J. Yuan, B. Sundén, Review on modeling development for multiscale chemical reactions coupled transport phenomena in solid oxide fuel cells. Appl. Energy **87**(5), 1461–1476 (2010)

110. J. Yuan, B. Sundén, On mechanisms and models of multi-component gas diffusion in porous structures of fuel cell electrodes. Int. J. Heat Mass Transf. **69**, 358–374 (2014)

111. R. Suwanwarangkul, E. Croiset, M.W. Fowler, P.L. Douglas, E. Entchev, M.A. Douglas, Performance comparison of Fick's, dusty-gas and Stefan-Maxwell models to predict the concentration overpotential of a SOFC anode. J. Power Sources **122**(1), 9–18 (2003)

112. S.H. Chan, K.A. Khor, Z.T. Xia, A complete polarization model of a solid oxide fuel cell and its sensitivity to the change of cell component thickness. J. Power Sources **93**(1), 130–140 (2001)

113. W. Lehnert, J. Meusinger, F. Thom, Modelling of gas transport phenomena in SOFC anodes. J. Power Sources **87**(1), 57–63 (2000)

114. Y. Patcharavorachot, A. Arpornwichanop, A. Chuachuensuk, Electrochemical study of a planar solid oxide fuel cell: role of support structures. J. Power Sources **177**(2), 254–261 (2008)

115. H. Zhu, R.J. Kee, A general mathematical model for analyzing the performance of fuel-cell membrane-electrode assemblies. J. Power Sources **117**(1), 61–74 (2003)
116. H. Yakabe, M. Hishinuma, M. Uratani et al., Evaluation and modeling of performance of anode-supported solid oxide fuel cell. J. Power Sources **86**(1), 423–431 (2000)
117. W. Li, Y. Shi, Y. Luo, N. Cai, Theoretical modeling of air electrode operating in SOFC mode and SOEC mode: the effects of microstructure and thickness. Int. J. Hydrogen Energy (2014 in press). doi:10.1016/j.ijhydene.2014.03.014

Chapter 5
Conclusions and Trajectories for the Future

From the earlier chapters, one can conclude that SOFCs are advantageous energy conversion sources mostly due to their remarkably high efficiency of converting chemical energy into electricity. Such achievable efficiency has driven various universities, research institutes, and companies to investigate the practical application of SOFCs in a number of areas, including automobile vehicles, self-powering homes, and wireless communication stations in rural regions. Toward the ultimate commercialization of SOFCs for everyday use, many aspects associated with SOFCs must be developed and refined further.

The relatively high working temperature, which typically requires heat-resistive and costly protection materials, resides as the fundamental challenge for the widespread application of SOFCs. The development of novel electrolyte materials and the rational design of heterostructured electrolyte materials are expected to help address the challenge. In particular, the introduction of lattice strains into the electrolyte crystals has shown pronounced efficacy toward enhancing the ionic conductivity of existing electrolyte crystals. Tensile lattice strain expands the lattice and allows for the ionic species to diffuse through the electrolyte crystals with lower activation energy. Nevertheless, several elusive scientific issues exist associated with the strain engineering of electrolyte crystals. First, the maximum tensile strain lies typically below 5 %, and dislocations are inevitably produced with the increased stain. Second, the existence of an analytical correlation between strain and ionic conductivity enhancement remains generally uncertain. Different groups have obtained dramatically different ionic conductivities even from electrolyte heterostructures with similar or even the same tensile strain. Third, lattice strain is typically dominant only in a limited number of atomic layers near the interface, and the diffusion mechanism of ionic species in strained electrolyte lattices remains in extensive, challenging investigation. Upon addressing these strain engineering issues associated with strain engineering SOFC electrolytes, the operating temperature is expected to decrease into a range that allows for the everyday commercial and industrial implementation.

The diffusion mechanism of fuel gas and oxygen species is more elusive in SOFCs with micro-/nanostructured electrodes. In such electrode materials, a few diffusion mechanisms can govern the gas transport simultaneously through the

© The Author(s) 2014
W. He et al., *Gas Transport in Solid Oxide Fuel Cells*, SpringerBriefs in Energy,
DOI 10.1007/978-3-319-09737-4_5

porous electrodes. The underlying scientific challenges include the fact that the analytical derivation of a comprehensive diffusion mechanism based on bulk diffusion and Knudsen diffusion is lacking. This leads to some uncertainty regarding diffusivity measurements that are conducted using currently available techniques. Therefore, much still remains to be explored to analyze the diffusion of fuel cell gas species with rationally designed/corrected diffusion models.

Then, a fundamental gap between developed diffusivity measurement techniques and actual achievable electrode materials seem to exist. Designing SOFC electrodes with efficient gas transport properties is but one step toward realizing functional electrode structures in experiments or in real world practice. Promising research trajectories can be devised to model and design SOFC electrodes from raw electrode materials and to ensure the feasibility of as-developed gas transport models toward the realization of efficient SOFC electrodes.

In the ideal case, electrolyte materials are single crystals that prohibit transport through the electrolyte crystals. In the real world, SOFC electrolyte materials can exhibit a certain extent of porosity; thus, gas transport through electrolyte between the two porous electrodes is possible. The transport of fuel cell gas species through the electrolytes reduces the voltage output of SOFCs and undoubtedly reduces the working efficiency of SOFCs converting chemical energy into electricity. Nevertheless, the gas tightness of SOFC electrolytes has remained a rarely explored subject until recently. Quantitative analysis over the gas tightness of intact SOFC devices is lacking in the field. Such investigations will allow for the quantitative evaluation of concentration polarization, ohmic polarization, and activation polarization associated with the operation of SOFCs. Upon addressing this measurement issue, one can design rationally the microstructure of the electrolyte materials and improve the overall performance of SOFCs for high-density power supply.

In parallel with gas transport, electronic conduction and ionic conduction are important forms of mass transport associated with SOFCs. The ideal operation of SOFCs is based on a closed electrical circuit composed of electronic conduction from anodes to cathodes and ionic conduction from cathodes to anodes through electrolytes. Therefore, the performance of SOFCs can be severely impeded by the inefficient ionic conduction in electrolytes. Understanding and improving the ionic transport through solid electrolytes rely highly on an efficient method of measuring ionic conductivity. A reliable measurement of ionic conductivity demands accurate analysis of ionic conduction and electronic conduction, both of which can contribute to the overall electrical conduction in solid electrolyte-based SOFCs. Electrochemical impedance spectroscopy (EIS), a commercially available technique, is the most commonly employed tool for measuring ionic conductivity in SOFCs. However, the validity of the ionic conductivity measurement through EIS can be debatable due to a few issues. First, the measurement system must be realized with materials that contribute none or little to the measured ionic conductivity. Efficient selection of materials can be particularly challenging, and special caution is typically required to extract the ionic conductivity of the electrolyte sample from the overall measured value. Second, electronic conduction often contributes to the overall measured electrical conduction; extracting ionic conductivity from the mixed AC impedance spectroscopy acquired within a range of frequency values can be challenging. Further, the ionic conductivity measurement *via* EIS is typically based on intact SOFC systems, which causes inevitably post-measurement waste of the electrolyte sample along with the other system components. Thus, an out-of-cell DC measurement methodology, eliminating the mixed conductivity contribution of electronic species, is highly desirable in the energy field. A series of simple low-cost DC-based electrochemical devices can be designed for the accurate ionic conductivity measurement in SOFCs. The measurement design and subsequent analysis on the measurement data will show unprecedented advantages over traditional measurement techniques. These devices are expected to find extensive interest and investment in both academic research and industrial production.

Looking toward the future, great opportunities are coupled with fundamental scientific challenges in the solid oxide fuel cell field. By addressing the theoretical and technological issues associated with the accurate measurement and evaluation of mass transport in solid oxide fuel cells electrodes and electrolytes, the high energy density and energy conversion efficiency will help mitigate the increasing social and economic unsustainability resulted from fossil energy resources.

Index

© The Author(s) 2014
W. He et al., *Gas Transport in Solid Oxide Fuel Cells*, SpringerBriefs in Energy,
DOI 10.1007/978-3-319-09737-4